For
a wise counsellor
and a good friend

5 iv 83

PARADOXES
OF
POWER

PARADOXES

OF

POWER

*The Military Establishment
in the Eighties*

Adam Yarmolinsky
and Gregory D. Foster

INDIANA UNIVERSITY PRESS • BLOOMINGTON

For J.

Blessed are the Peacemakers

Manufactured in the United States of America

Library of Congress Cataloging in Publication Data

Yarmolinsky, Adam.
Paradoxes of power.

Includes index.
1. United States—Defenses. 2. United States—Armed
Forces. I. Foster, Gregory D. II. Title.
UA23.Y29 1983 355'.033073 82-48523
ISBN 0-253-34291-0
1 2 3 4 5 87 86 85 84 83

CONTENTS

Preface

This is a second look at an institution I first looked at in detail more than a decade ago in *The Military Establishment,* after spending some six years working at or near the center of that establishment, and four years observing it from academia. Since then the establishment's situation has shifted, as has my perspective. The result is a book that borrows only a few numbers from the old text.

I am particularly grateful to my coauthor, Gregory Foster, whom I was fortunate to be able to recruit at a critical stage in the project. He has provided essential facts and ideas, on an extraordinarily tight timetable.

My thanks go also to Ron McMahon, an early and helpful participant; to Jim Skelly, who furnished valuable information and insights to the chapter on social impacts; and to Paul Bracken, Barry Carter, Paul Doty, Robert Kupperman, Jack Ruina, John Steinbrunner, Jack Stockfisch, and Paul Warnke. All of them contributed in various ways, directly and indirectly, to my understanding of the subject matter, but none of them is in any way responsible for the result. My secretaries, Joan Chevalier, Chandanie Froemming, and Dorothy Martin have endured many retypings.

Lastly, I am grateful to my children, who continue to inform my vision of the world in which all this takes place.

A.Y.

Washington, D.C.
1 June 1982

1

The Elements of the Problem

THE UNITED STATES military establishment is important to many different kinds of people in many different ways. It is a way of life for some two million men and women who serve in it as soldiers, sailors, airmen, and marines, whether they are recruits on a first enlistment, anxious to get it over with, or "lifers" serving out their 20 or 30 years to retirement. It is a paycheck for another million Americans, about half of them blue-collar, who work for the Department of Defense the way other people work for the Postal Service; and for several million more who work for defense contractors. It is alternately a (limited) bonanza and a dry hole for those companies that depend heavily on defense business and that compete, although not in classical economic fashion, for multibillion dollar defense contracts. It is Big Daddy for army (or navy or air force) towns across the country. For the industrial strategist, whether in the planning department of a great corporation or in the halls of government, the establishment represents a major claim on national resources, primarily in that it diverts research and development away from more readily marketable products and sops up technical personnel who may be in short supply. It is the biggest budgetary headache for the president and members of Congress; yet almost all of them agree that it needs more dollars than it is getting right now, even if very few of them agree on how much more or on whom those dollars should be spent.

This enormous structure is no monolith of power. It engages in constant and often intense self-examination. Within the past year, the retiring chairman of the Joint Chiefs of Staff and the Chief of Staff of the Army have proposed major changes in the central organization. Interservice rivalry is an ancient phenomenon in the military and results in sharp and often violent criticism of one service by another.

But the massive size and pervasive influence of the military establishment throughout American society and its role as at least one element in whatever power and influence the United States projects beyond its borders make the establishment a fit subject for study by every citizen.

This is not to suggest that the United States is a garrison state or is in serious danger of becoming one. (Even if President Reagan's unprecedented increases in the proposed defense budget through 1987 are approved by the Congress, the increase in the military establishment's share of the gross national product will be just over 1 percent.) Rather, the establishment is so heavily and extensively involved in almost every aspect of American life that we cannot understand our own society without learning something about the anatomy and the physiology of this elemental component and its intimate interaction with the civilian world.

It is a principle of political as well as physical dynamics that impact is a function of mass; the greater the mass the greater the impact. But impact is also a function of speed and direction. Even over its recent history, the United States military establishment has experienced such major changes, both in its growth rate and in its direction and purpose, that assessing its internal and external impacts becomes an extremely complex enterprise. The United States military was still concerned with "pacification" of native Americans only a century ago. It felt wholly secure behind a two-ocean barrier only half a century ago. It entered the nuclear age (along with the rest of us) less than two generations ago. It is only just recovering from losing a war for the first time since 1812, and it has already been told to plan for an expansion in spending, if not in numbers, greater than the expansion for the Vietnam War. And even a vault full of contingency plans cannot tell the military what its civilian masters expect of it tomorrow.

Further, the component elements of the military establishment—the four military services, the civilian authority, the competing industrial and regional interests, the differing factions within the research and development and intelligence communities—are often in pursuit of quite different bureaucratic objectives, so that the institution that may appear monolithic to the distant lay observer is in fact a cave of the winds.

At this writing, the major concern of the American people about the United States military establishment is whether it is big enough to meet its responsibilities, and particularly to face the challenges that seem to be generated in various parts of the world, mostly outside the continent of Europe, by the other superpower. Whether the Soviet Union is acting out an aggressive design or is reacting with paranoia and opportunism, the combination of its growing military strength and the increasing fragility of existing regimes in so many parts of the world has fueled traditional anxieties about the adequacy of our military strength in maintaining some kind of global equilibrium.

At the same time, however, another strain in American thought and American politics is the continuing concern about the fact that the United States military establishment is still the largest operating organization in the country. President Eisenhower expressed his frustration, as he left office, with the "military-industrial complex"—perhaps his most quoted phrase. The new Department of Health and Human Services, even stripped of its functions in education, has a bigger budget than the Department of Defense because it makes such enormous transfer payments. Nevertheless, it is a much smaller operation. In fiscal 1982, defense accounted for some 70 percent of total federal procurement (what the government buys) and over three-quarters of the federal payroll. The military establishment remains almost as visible, if not as vulnerable, to its critics as it was in the seventies.

There are four major paradoxes that surround the United States military today:

To begin with, there is the paradox of deterrence: the only way to avoid the ultimate conflict—general nuclear war—is to be prepared to respond with overwhelming force to an attack against which there is no adequate defense. We must deter because we cannot defend, in a war that no one can win. If we are ever forced to employ our nuclear weapons, both we and our adversaries will have lost.

Then there is the paradox of limited response: the stronger a great power the more careful it must be to limit its military objectives so as not to escalate to general nuclear war. A small country, an Israel or a Syria, even a Britain or a France, can afford to take risks or embark on military adventures with reasonable confidence that it will not be precipitating a nuclear Armageddon.

Third, there is the paradox of military bureaucracy: the size and complexity of the military establishment that a superpower must maintain in order to preserve an international power equilibrium may be beyond the capacity of a national government to manage or to contain within the reasonable limits of domestic political power struggles. No coalition of civilian agencies can come close to matching the Defense Department's resources. If the number of military and civilian personnel were cut in half, the department would still have almost one and a half times the total number of employees in the rest of the Executive Branch.

Finally, there is the paradox of peace: How can a nation purportedly dedicated to peaceful purposes live with a large military establishment that plays a predominant role in national affairs? But on the other hand, how can it live in the world today without such an establishment?

All these paradoxes are increasingly complicated by the fragmentation of military and political power throughout the world, among fractionating nation-states and even among private groups outside the law.

As the military establishment is both a generator and a victim of these paradoxes, so it both exerts a powerful force in the international and domestic affairs of the United States and reacts to other forces, foreign and domestic, that have shaped its past and will shape its future.

We begin with an analysis of the conflicting demands placed on the military by its civilian masters, and of the structure of the military establishment, within and beyond the Pentagon. We then examine the military in the American economy, and in the American polity and American society at large. We return to the theme of civilian management and control as it has developed over the life of the Republic. And we conclude with a look into the clouded glass, at alternative futures for the military (and for the world) and lastly at arms control as a key determinant of that future—and of whether we survive to enjoy it.

But before turning to this analysis, we need to examine briefly the major factors that affect the thinking of various groups in American society about the military establishment, about the assignments that we give to the establishment, and about the resources that we make available to it in order to carry out those assignments.

The single most powerful determinant of attitudes toward the military establishment over the past third of the century has been the state of tension between the United States and the Soviet Union, and, not entirely as a complement to that tension, the global responsibilities and ambitions of the United States.

At any one time, both these factors are very much a function of what has happened up to that point. What looks to some American leaders like a Soviet game of catch-up appears to others as a plan for military domination. The Committee on the Present Danger, organized in 1976, had its counterpart in a committee of the same name organized in the United States in 1950. Although there are important differences between the two committees, there is nonetheless a striking similarity in outlook as demonstrated by the following pronouncements: "The aggressive designs of the Soviet Union are unmistakably plain." "American power is far from adequately mobilized."* The critical historical parallel for some American leaders

*A number of highly placed officials in the Reagan administration served on the board of directors of the current committee. These include William Casey, director of the Central Intelligence Agency; Fred Ikle, under secretary of defense for policy; Jeane Kirkpatrick, United States representative to the United Nations; John Lehman, secretary of the navy; Paul Nitze, chief negotiator for the Intermediate Range Nuclear Force Talks; Eugene Rostow, director of the Arms Control and Disarmament Agency; and Richard Allen, former assistant to the president for national security affairs.

during the Vietnam buildup was the appeasement at Munich; for
others it was the dangerous extension of the land war in Korea.
Soviet motives are more difficult to fathom, but surely the enormous Soviet weapons buildup of the seventies was at least in part a
reaction to the humiliation of the Cuban missile crisis, which in turn
may have been in part a reaction to the frustration of Soviet ambitions for the removal of Western power in Berlin.

In any event, the two major factors that seemed to play off
against each other in recent attitudes toward the military are, on the
one hand, the mixture of caution and some degree of shame over
United States involvement in Vietnam, and, on the other hand, a
pervasive if not precisely definable fear that United States power in
the world has been declining, not only vis-à-vis the Soviets, but also
in our dealings with other countries, from Latin America to the
Near East, to Africa, and to South Asia. Whatever the objective
facts about United States military power, we are dealing here with
attitudes and perceptions that combine a good deal of ambivalence
and hesitation about the uses of military force in the world with a
good deal of worry about the adequacy of existing and projected
forces over the decade of the eighties. There is a strong feeling,
justified or not, that somehow the United States has been pushed
around in recent years more than a great power ought to be.

These conflicting concerns are further exacerbated by two
looming domestic shortages: a chronic shortage of money, and a
current shortage of manpower. Clearly the military establishment is
not being shortchanged in the budget projected over the next four
years, even though the defense budget has not retained its status as
the largest piece of the federal spending pie. Indeed there is a
serious question whether the Defense Department will be able to
spend intelligently all the dollars that are being pushed in its direction. But the department's immunity to budgetary pressures is not
guaranteed, even to 1984. The defense budget remains one of the
few budget items for which there is at least some apparent flexibility. So much of the federal budget consists of fixed items of spending, many of them mandated by law, that competition between
discretionary domestic spending and discretionary defense spend-

ing is inevitable, all the more so as federal spending is claimed by the Reagan administration to be the major factor in inflation.

In the present climate, however, the shortage of recruits from the shrinking pool of 18-year-olds tends to make people a more critical resource for the military than dollars. The problem is only temporarily alleviated by the effect of the economic depression manifested in particularly high unemployment among young people. In the light of the political difficulties in reinstituting a peacetime draft, the military establishment may soon be faced with a choice of joining in a major national effort to popularize general voluntary service, both civilian and military—or of making do with a total force significantly smaller and of lower quality.

Among major factors likely to shape the military establishment over the next decade, one cannot exclude arms control. In 1977 or 1978 this observation would have sounded like an absurdly cautious understatement. The Strategic Arms Limitation Talks between the United States and the Soviet Union were moving toward an agreement that was expected to stop the seesaw strategic nuclear arms competition and even begin to turn it around. At least six other significant negotiations were under way: the comprehensive test ban talks, the antisatellite talks, the Indian Ocean arms limitation talks, talks on chemical warfare and so-called radiological warfare, conventional arms sales talks between the United States and the Soviet Union, and even the so-called Mutual and Balanced Force Reduction talks between members of NATO and the Warsaw Pact, which had been moving at a majestic if not somnolent pace for a number of years, but were showing signs or at least possibilities for more lively development. It may be instructive to recall how recently this encouraging atmosphere prevailed, in the light of current discouragement about foreseeable prospects, at this writing, for the resumption of constructive talks about strategic arms limitations with the Soviet Union and the Eastern bloc. So long as the United States treats the resumption of negotiations as a favor to the Soviets, in return for their (relatively) good behavior, it seems questionable whether the current conversations will be serious ones. But the costs and the dangers of mindless competition in the production and refinement of weapons of mass destruction, as well as pressures from an aroused American public and from allies who see negotiations as a necessary counterpoint to modernization, suggest that the tide must turn.

There is a serious question as to whether arms control agreements, even in the most favorable atmosphere, can keep pace with competition in new, more deadly, and perhaps more destabilizing arms technology. But even with the slower and more hesitant pace of negotiations, it seems incontrovertible that arms control considerations are entering into the choice and the development of new weapon systems. These considerations are involved not only in attempts to stop or reverse the growth of nuclear forces but also in decisions about force deployments and rules of engagement.

It has not been so long since the chairman of the Joint Chiefs of Staff wondered aloud at the inconsistency of pursuing simultaneously efforts to make up shortfalls in our military arsenal and arms control initiatives. Nowadays the military takes it for granted that it cannot discuss new weapon systems or new deployments without considering the arms control implications of each option—and of their interactions with each other. Today, it is the civilian political leadership that seems to favor all-out arms competition.

All these factors sharpen the paradoxes of decision making, for a superpower that is also a democracy, about the size and shape of its military establishment. The country responds to intentions of our principal antagonists (and of lesser ones as well); the needs, hopes, and beliefs of our friends; the residue of previous national traumas; the limitations imposed by budget shortages in a world beset by inflation and by people shortages on the down side of the birthrate cycle; and heightened awareness of arms control considerations.

II

What the United States Expects of Its Military Establishment in Peace and War

AMERICANS ARE DISCOVERING more and more things that the United States military establishment cannot accomplish. The list begins with the realization, sometime in the 1960s, that no military establishment can defend against a nuclear attack. It can only hope to discourage the potential attacker by threatening devastating retaliation. The United States military establishment did not succeed in a massive intervention in a guerrilla war in Southeast Asia. Whether it might have succeeded by employing different tactics is at least doubtful, but the fact remains that committing 500,000 Americans, with massive armament and equipment, did not produce victory. The United States military establishment cannot always protect individual Americans, even when they are hedged about with diplomatic immunity, as was amply demonstrated by the ordeal of the hostages in Iran and the ill-starred rescue attempt. Nor could the United States military restore the independence of Afghanistan without incurring unacceptable risks—risks that are a consequence of the overhanging threat of nuclear Armageddon, not of United States military strength or weakness.

What then are the functions of the United States military, as they are generally understood, and where do they shade off into confusion and conflicting exceptions? The first task of the U.S. military is to prevent the ultimate catastrophe of general nuclear war. It tries to do so by maintaining a nuclear arsenal powerful enough and secure enough to withstand the most devastating attack that could be launched against it and still be able to strike back to inflict unacceptable damage on the attacker. As the great philosopher of nuclear strategy, Bernard Brodie, put it, "Thus far, the chief purpose of our military establishment has been to win wars. From now on, its chief purpose must be to avert them." Some of the arguments about the elements of the deterrent strategy are explored below, but the centrality of the deterrent function remains unquestioned.

The second mission of the United States military establishment is to help protect our friends and allies against attack and to give them the confidence to maintain independent policies, even in the face of bullying by another superpower—or by a smaller antagonist. This protection is not necessarily a matter of extending the famous United States nuclear umbrella over the territory of our allies; it is more a matter of maintaining the kinds of nonnuclear forces— trained, equipped, and deployed, with adequate prepositioned supplies—to put on the attacker the terrible burden of deciding whether to escalate to nuclear weapons or to withdraw.

The third mission of the military establishment blends indistinguishably into the arts of diplomacy and of police work. This is the function of deterring and defending against low-level violence and international terrorism. Inside the United States, the military role is a sharply limited one, supplementing civilian police forces—and the National Guard units mobilized under state authority—only in extreme emergency. Outside the United States, the need for military help is more likely to develop, but providing it is an even more delicate matter. While equipment, advice, and training may be all that is necessary or appropriate in the way of United States assistance to a country, there is always the danger of being drawn into deeper involvement, including the active military support of a regime that has lost the confidence of its own people.

There is yet a fourth function performed by the military. Perhaps inadvertently, even unwittingly, the military establishment is an institution of social reform. As such, it cannot ignore such social problems as poverty, racial unrest, drug abuse, crime, and general alienation, but must cope with them within its own ranks. This is particularly true in an all-volunteer environment, where manpower is at such a premium. As the economy deteriorates, more and more out-of-work and untrained youth turn to the military. One is reminded of the army recruiting slogan: "We don't ask for experience; we give it." To some extent, therefore, the military must face the task of making successful soldiers out of young people who have failed at everything they have tried up to that point.

Lastly, the United States military establishment needs to be prepared as best it can to protect the United States from unforeseen dangers and to defend American interests that may be threatened in places and fashions that we have not yet imagined.

Defending America's frontiers is not a significant responsibility of United States armed forces. We are fortunate in our neighbors and in the protection that two oceans afford against nonnuclear attack. Defending the sea lanes of United States commerce is another matter, however, and a particularly troublesome one, as dependence on foreign oil for almost 40 percent of our supply and 16 percent of overall United States energy needs continues as the most serious vulnerability of the economy and the society.* Here again the issue is more than one of simple defense. The United States fleet cannot by itself guarantee the safety of the United States Merchant Marine (such as it is), much less of the foreign-flag ships that carry the great bulk of United States foreign commerce and 97 percent of our oceanborne oil supplies. But to interfere with the foreign commerce of the United States would be an act of war, and therefore all the elements of the United States military establishment taken together serve as a deterrent to attacks on that com-

*It is worth noting, however, that the salt caverns in Louisiana and Texas, which constitute our Strategic Petroleum Reserve, now hold approximately 225 million barrels of oil. Some experts have argued that, when this reserve is filled to its presently projected 750 million barrel capacity, the United States will be able to absorb a daily loss of 2 million barrels of imported oil for a year.

merce. What this means is that particular missions can no longer be identified with particular elements of military force—if indeed they ever could be in the past. The advent of nuclear weapons underlies the critical importance of deterrence and, by the same token, tends to decouple forces from specific political missions.

In order to accomplish the first mission, prevention of nuclear war, the United States maintains a so-called triad of strategic forces, including long-range and medium-range bombers, land-based missiles (ICBMs) and submarine-launched missiles (SLBMs)

Concern (some would argue unjustified concern) about the age of the 300-odd B-52 bombers has produced plans to replace them with a new bomber, the B-1, which should be ready, with luck, to enter the force in the late 1980s, but will only be able to penetrate Soviet defenses until some time in the 1990s. At that time it is proposed that it be succeeded by a bomber labeled "Stealth" because of its very small radar profile, making it much less vulnerable to antiaircraft weapons.

The 1,000 Minuteman missiles (plus 54 older Titan-II missiles) make up the ICBM force. The greatly increased accuracy of Soviet missiles would arguably make 90 percent or more of the United States missiles vulnerable to a first-strike attack sometime during the 1980s, creating what the Reagan administration has labeled a "window of vulnerability." A new (and itself more accurate) missile, the MX, is being built to replace the Minuteman. The administration has not, however, been able to settle on a way of deploying the new missile that will make it relatively invulnerable to a Soviet attack and that will be politically acceptable. Various deceptive basing modes have been rejected on technical grounds and because they are unacceptable in the states where they would be deployed. Interim plans to place the MX in Minuteman silos with additional hardening of the silos have been severely criticized as an inadequate makeshift. Meanwhile, critics point out that no sane Soviet planner would risk an attack on the Minuteman missile force, given the extraordinary range of technical uncertainties involved; the inevitability of reprisal from manned bombers; and the submarine-launched missile force, which remains, and will remain for the foreseeable future, invulnerable to attack. The SLBM force, consisting of some 650 missiles in 41 submarines, is being substantially modernized. Most United States missiles are equipped with multiple independently targeted warheads (MIRVs),* and the United States has some 10,000 strategic warheads, as compared with the current Soviet total of some 8,000.

The second mission, of protecting friends and allies against attack and bullying, is the primary responsibility of the army, navy, marines, and air force units stationed around the world and reserves within the United States. Of the more than three quarters of a million men and women in the active duty army, over 200,000 are in Europe and 30,000 in South Korea. The Marine Corps maintains one of its three divisions and one of its three air wings in Japan. The navy is organized into four fleets—in the Atlantic, the Mediterranean, the eastern Pacific, and the western Pacific—while the air force has major units stationed in Europe and the Pacific, as well as a Tactical Air Command and a Military Airlift Command based in the United States. A Rapid Deployment Force, designed particularly for contingencies in the Middle East, is being organized, but will not be deployed until the late 1980s.

Another way to look at the obligations of the United States military—and the expectations of the American people—is continent by continent and area by area, although here again the United States response to a threat in one part of the world may best be effected in a quite different part.

Currently, we have about half a million troops dispersed around the globe. These consist of a third of a million in Europe; 135,000 in the Pacific and the Far East; 15,000 in Latin America; and the remaining 25,000 in various other overseas locations (including nearly 21,000 naval forces afloat).

Over two generations, from 1940 to 1980, the proposition was scarcely debatable that any tour d'horizon of United States security problems and military commitments around the world necessarily

---

*Multiple Independently-Targetable Re-entry Vehicle, a package of two or more nuclear warheads that can be carried by a single ballistic missile and delivered on separate targets.

begins in Western Europe. Even the shooting wars in Korea and Vietnam were sideshows, far from the main arena of potential conflict. With the beginning of the new decade, this assumption has been called into question by three new, or at least newly visible, developments. The first is the increasingly evident vulnerability of the United States supplies of raw materials from other parts of the world, particularly Middle Eastern oil—and the even greater vulnerability of our European and Asian friends and allies.* The second is the increasing instability of Central American regimes and the concern that the Soviet Union, primarily through its Cuban surrogate, may be exploiting these instabilities in ways that could somehow threaten the security of the United States. The third, and most disturbing, development is the widening rift within NATO itself, with at least the suggestion that the United States may pull back from its long-standing commitments in Europe and either enter a new isolationist phase or reorient itself toward the Pacific.

Nevertheless, we begin with the obligations of the United States to its NATO allies, enshrined in the 1949 treaty and reaffirmed in 1969. These obligations assume that in the highly improbable but not inconceivable event that Allied forces are attacked and overrun by the armies of the Soviet bloc, the United States, with the consent of its allies, would be prepared to initiate the use of tactical nuclear weapons, at least at the battlefield level. Or, in the even more unlikely event that the Soviets were to initiate a combined attack with nuclear and nonnuclear forces, the United States, again with the consent of its allies, would be prepared to respond in kind.

Whether the United States and its allies could resist a Soviet bloc attack with conventional forces, and for how long, is a question

---

*The dependency of the United States on foreign sources of nonfuel minerals and metals has increased markedly over the last two decades. Of the top 25 such imported commodities, in 1960, our dependency averaged 54 percent. In 1980, our dependency for the same items averaged 70 percent. In fact, our dependency is 75 percent or more on foreign countries where war could, in the foreseeable future, deny us our supplies of bauxite (93 percent), chromite (91 percent), cobalt (95 percent), columbium (98 percent), manganese (97 percent), nickel (76 percent), and tantalum (98 percent).

on which experts differ. They even differ as to whether in a general war the Soviets would initiate the use of nuclear force at the same time that they attacked with conventional force. The weight of expert opinion would appear to be on the side of Soviet conventional (i.e., nonnuclear) superiority, based on the Warsaw Pact's advantage in numbers of tanks (3 to 1) and planes (2 to 1) although NATO has more antitank weapons and helicopters and a slight edge in manpower (1.4 to 1), at least in the initial stages of the conflict. Some qualified observers would argue that NATO's qualitative superiority in equipment, taken together with more extensive and perhaps more sophisticated training, would shift the balance.

But without attempting to resolve this controversy, one can note the consensus of popular views and political leaders on both sides of the Atlantic—a consensus maintained for more than a generation—that it is the United States nuclear deterrent that protects the alliance, and that the function of American troops on the ground in Europe is first to convince the Soviets that they cannot take military action in Europe without involving the United States directly, and second to gain some time for the Soviets to ponder the consequences of continuing to advance.*

This consensus has been remarkably persistent. At times there has been somewhat more optimism about the capability of the Allied forces to hold back a bloc attack without invoking nuclear weapons. At times there have been doubts, particularly in Europe, about the willingness of the United States to put its cities at risk, by launching a nuclear attack that could destroy Russian cities in response to a successful massive Soviet attack on Western Europe, with or without nuclear weapons. President Reagan's November 1981 public statement to the effect that a tactical nuclear exchange could be confined to the battlefield has done little to assuage European fears about United States resolve.

These doubts have had two consequences: The first is that some Europeans have asked for nuclear weapons capable of reach-

---

*The consensus has recently been challenged by four prominent Americans who propose that the United States adopt a policy of no first use of nuclear weapons. See chapter 8.

ing the Soviet Union itself, based on European soil—although the logic of the demand is questionable, since these weapons would still be controlled entirely by the United States—while other Europeans have expressed vigorous objections to any nuclear weapons on their soil (this point is discussed in greater detail in chapter 8 below).

At the present time, the Soviets have a decided and growing edge in so-called theater nuclear weapons, specifically the SS-20 mobile missiles, which can reach targets throughout Western Europe.

The United States proposes to deploy two kinds of missiles to meet this threat: so-called extended-range Pershing missiles, tactical ballistic missiles able to reach Soviet territory; and cruise missiles, which are in effect small pilotless airplanes, capable of navigating to preassigned targets within the Soviet Union. Both kinds of missiles would have nuclear warheads and would be entirely under United States control.

European members of NATO are torn between a concern that the imbalance be rectified, in part so that negotiations for reductions on both sides can proceed on an equal basis, and a complementary concern that overemphasis on a so-called theater balance in nuclear weapons may tend to "decouple" the United States strategy calling for a nuclear response from United States-based bombers and missiles if a nuclear attack is launched on Western Europe. Put in oversimplified form, as many Europeans are inclined to do, Europe is afraid of becoming the nuclear battleground in a war between the Soviet Union and the United States.

The second consequence of doubts about United States resolve is that some Europeans, and some Americans, have expressed the view that effective deterrence depends on the ability of United States nuclear forces to strike back at Soviet military targets rather than at Russian cities. During the 1980 presidential campaign, there was much talk about a new presidential directive, PD 59 (the text of which is still secret), which was said to abandon the doctrine of mutual assured destruction for a strategy of pinpointed retaliatory strikes on Soviet missile sites and other military targets. It was even charged that the United States was adopting a "nuclear war-fighting doctrine," although the charge was vigorously denied by the secretary of state (who had learned about the directive only after it was leaked to the press). The fact is that the United States has been pursuing a modified European strategy since 1961, when it was first articulated by Secretary of Defense Robert S. McNamara, in his Ann Arbor speech, permitting a selective response to less than all-out nuclear attacks.* While PD 59 was described by Secretary of Defense Harold Brown as only a codification of existing doctrine, the modified doctrine, taken together with the relentless development of more and more accurate nuclear weapons, tends to create a situation in which what one side sees as a capacity for limited retaliation against military targets, the other side may see as an attempt to obtain a first-strike capability. If the Soviets believe that we can destroy a large portion of their nuclear forces, 70 percent of which are still in relatively exposed land-based missiles, they may be tempted in a crisis to undertake a preemptive strike.

A more troubling development is the number of suggestions emanating particularly from Secretary of Defense Weinberger, that at least one part of the Reagan administration contemplates the possibility of engaging in protracted nuclear warfare. The thought of isolated missile-launching crews exchanging salvos in a world otherwise completely devastated is reminiscent, on a global scale, of the unit commander in Vietnam who explained that he had to destroy a village in order to save it.

In the European situation, there has been an increasing disjunction between European and American attitudes. The European nuclear disarmament movement attracts increasing support in half a dozen Western European countries. If, as seems quite possible, there is no significant progress in the theater nuclear weapons arms control talks between the Soviet Union and the United States before the present timetable calls for deployment of new weapons in Europe, European governments may be unwilling or unable to accept them, and irresistible pressures may be generated in Congress to withdraw American troops, because, it may be argued (however

---

*NATO did not adopt "flexible response" until 1967, and then reluctantly.

illogically) they are not adequately protected in the event of nuclear escalation. The same controversy could arise on the issue of deployment of the so-called neutron bomb, a tactical nuclear weapon using enhanced radiation that it is claimed will kill soldiers—primarily tank crews—without as much devastation of the surrounding area as would result from the use of ordinary tactical nuclear weapons.

These differences, although fundamental, could be resolved. The principles of common defense embodied in the initial treaty are as sound as when they were first agreed to. The members of NATO have the collective strength and capacity to organize themselves to meet the Soviet challenge without resorting to nuclear weapons— although the extraordinary complexity of organizing a collective alliance defense in Europe gets too little attention outside military circles. The difficulties of achieving what the military calls "interoperability" of military units and their equipment appear to be a subject of interest only to the experts.*

But the significant fact for the United States military is that it is being asked to prepare to deploy in Europe weapons that may be unacceptable to its European allies, while it is still a long way from developing the kind of forces that could require a Soviet aggressor to make the terrible decision whether to escalate to nuclear weapons or to abandon its aggression. Instead, the United States military may find itself, in a few years, in a situation where it may be under great political pressure to bring the troops home from Europe because they have been unable to deploy nuclear weapons that they don't really need—but that are seen in the United States as essential for their protection.

It has been an axiom of military force planning for at least 20 years that "Europe drives the force structure," i.e., that the requirements for troops and weapons to meet United States obligations within NATO determine the size and shape of the overall United States military force structure. The thickening atmosphere of confu-

---

*"Interoperability," and the companion concepts of "rationalization" and "standardization," have been part of the NATO lexicon for several years now and are the subject of an annual report that must be submitted by the secretary of defense to Congress.

sion and confrontation over NATO military requirements makes it harder for the United States military to prepare for what may be expected of them in the rest of the world as well.

Moving eastward, the Persian Gulf and its surrounding oil-rich (and war-torn) territories have been the major focus of recent concern over relative United States and Soviet military capabilities. Again, United States military capabilities here are seen as limited by geography and by the inherent difficulty in keeping the oil fields, the pipelines, and the sea-lanes open if threatened with harassment or shutdown. The United States is now in the process of acquiring nearby facilities to preposition equipment and stage forces (we eschew the word "bases") and is developing a Rapid Deployment Force, but one that at its eventual size (200,000 active troops plus up to 100,000 reservists) and composition may not be fully deployable until the end of the present decade. Still, it is the president's effort to communicate to the Soviets that the United States has a vital national interest in maintaining the flow of oil from the Middle East to itself and to its allies that undergirds the ultimate threat of nuclear war to protect those installations from Soviet intervention. What the United States might be prepared to do to assist the oil-producing countries themselves in overcoming internal violence or disruption remains necessarily vague, particularly since such operations would have to be conducted in the Soviet Union's backyard.

The military situation in the Persian Gulf is further complicated by United States commitments to the state of Israel, which is surrounded by hostile neighbors, some of them supported by Soviet arms. The Israelis insist they can defend themselves with their own man- and woman-power, but would the United States stand by if Israel were being pushed into the sea? And if American forces were involved in so troubled an area so close to the Soviet Union, what consequences might result? On the other hand, an explicit guarantee by the United States of Israel's national existence (perhaps conditional on abandonment of Israel's expansionist policies) might defuse the potential for renewed all-out war. In any event, the continuing formal state of war between Israel and the Arab countries makes United States contingency planning for the area even

more subject to uncertainties—as does the shifting balance within
the Arab world. In the Iran-Iraq war, for example, the United States
was first concerned that Iran might lose, and later that it might win.

South of the Gulf lies the Indian Ocean, where earlier hopes to
hold down or even to cut back military use have given way to efforts
to increase the United States naval presence, primarily to protect oil
shipping lanes and to be poised for unspecified activity in and
around the Persian Gulf.

The British-owned island of Diego Garcia is rapidly becoming a
warehouse for United States activities in the region. Although
located over 2,400 nautical miles from the Strait of Hormuz—the
entrance to the Persian Gulf—the island can provide support for the
two aircraft battle groups on rotational duty in the Arabian Sea.
Runways are being enlarged to make it easier to handle long-range
B-52 bombers. The United States has about 1,500 troops stationed
on Diego Garcia, and the total investment there could approach $1
billion. Other agreements have provided the United States access to
Oman's Thamarit airfield and to its air and naval facilities on the
island of Al Masirah, to Kenya's port of Mombasa, and to the Egyp-
tian military base at Ras Banas. Our naval facilities at Berbera in
Somalia on the Horn of Africa, just down the coast from Soviet naval
installations in Ethiopia, threaten to involve the United States and
the Soviet Union in the continuing conflict between the two African
neighbors.

On the other side of the African continent, the United States
has been sparring with Muamar Quadaffi's Libyan regime, which
directly threatens its North African neighbors and apparently sup-
ports terrorist activity around the globe. Meanwhile United States
military aid has been resumed to Morocco in its war with the Alge-
rian- and Libyan-backed Polisario, over a barren stretch of the western Sahara. At the southern end of the continent, Cuban soldiers
support the Soviet-oriented regime in Angola, where their duties
apparently include guarding the facilities of American oil companies
against guerrillas who may or may not be financed by the Central
Intelligence Agency. Neighboring South Africa, feeling itself increasingly isolated because of its racial policies, proclaims itself a

bulwark against Communism and presses for United States military assistance. United States military forces are highly unlikely ever to be involved—but the possibility remains.

Beyond the Middle East, Soviet forces in Afghanistan are seen by some as a potential threat to neighboring Pakistan, and, through Pakistan, again to the Persian Gulf. Since the Soviet invasion of Afghanistan, the United States has been providing military assistance, overt and perhaps also covert, to the not-altogether-stable Pakistani government, despite continuing concerns about possible development by Pakistan of its own nuclear weapons. If Pakistan were threatened with imminent invasion—or if Soviet troops actually crossed the Afghanistan-Pakistan frontier—would United States forces be rushed to the scene, at least as a symbol of American commitment? And how would United States military involvement in Pakistan be squared with the need to maintain traditional friendly, if somewhat distanced, relations with India, Pakistan's antagonistic and much larger neighbor to the south?

Still farther east, both Japan and Korea are covered by the United States nuclear umbrella, and American ground troops share, at least symbolically, the responsibility for guarding the border between South and North Korea. Even while Japan contemplates an expanded (and much more expensive) military establishment, it depends on the United States to keep open its sea-lanes, on which it relies for essential oil and other strategic supplies. Efforts by the United States government to badger or cajole the Japanese into significant increases in their military budget have been largely unsuccessful and have resulted in some difficulties within the Japanese government. Plans to phase out United States ground troops in Korea have been indefinitely postponed, as a harsher South Korean military dictatorship and a more strident North Korean Communist regime threaten the fragile truce between the two halves of the peninsula.

With the effective dissolution of the Southeast Asia Treaty Organization (SEATO) in 1975, the United States has no formal treaty obligations per se remaining in Southeast Asia. If, however, the Soviet-backed Vietnam government were to attack Thailand across the Cambodian border, and Thai forces were to flee in retreat, the United States could be moved to provide more than supplies and spare parts to support the one country in the region that has traditionally maintained its independence both from the colonial powers and from its neighbors. The United States commitment to Thailand is embodied in the 1962 Rusk-Thanat communiqué and in the SEATO treaty itself, which officially still remains in effect.

Communist China has been seen as a potential military threat to the United States since the heyday of the China lobby in the 1950s. When China first began to develop its own nuclear weapons, some analysts even discussed the possibility of a preemptive nuclear strike against Chinese nuclear facilities. Even after the Sino-Soviet split in 1957, the United States continued to be concerned about Chinese military involvement in Southeast Asia and on the northern borders of the Indian subcontinent, and the possibility of a military confrontation with the Chinese was a major inhibition on the United States conduct of the Vietnam War. It remained an issue until the United States withdrew from Vietnam. With normalization of relations early in 1979, the situation was changed dramatically. At first, American policymakers were convinced that any attempt to play the China card against the Soviet Union would only amount to helping China play its American card, and that risks to world peace would outweigh any possible benefits. The principle that the United States should do nothing to exacerbate Sino-Soviet differences has been somewhat eroded by the events of 1980 in Poland and Afghanistan. At the same time, the United States continued to trouble the waters in its relationship with mainland China because of the Reagan administration's autumnal romance with Taiwan, a rapprochement reinforced by at least part of President Reagan's own party.

Completing the circuit of the globe back to the Western Hemisphere, the United States commitment to the independence of its southern neighbors from extrahemispheric pressures is our oldest foreign policy, embodied in the Monroe Doctrine. At this writing, the Reagan administration has dispatched advisors to assist the military forces of the elected government in El Salvador against an insurgency fueled by a mixture of Marxist doctrine and aroused resentment over centuries of oppression. Paradoxically, the same military forces (or at least an important element) are attempting to restrain the rightist tendency of the elected government, where a coalition of minority right-wing parties has so far denied power to the centrist Christian Democrat Party, which won a plurality but not a majority in the recent election. In neighboring Nicaragua, an avowedly Marxist regime is apparently funneling supplies from Cuba to the Salvadoran rebels, while still seeking an uneasy accommodation with the United States. Other El Salvadors—or Nicaraguas—may emerge from bloody and repressive regimes in other Central American countries.

Meanwhile, off the southern tip of the continent, Argentina seized and briefly held the bleak moors of the Falkland Islands, to which they have long laid claim, and the British navy rushed, somewhat belatedly, to the defense of the island population, which had no desire to be incorporated into Argentina. The United States tried in vain to mediate but came down almost inevitably on the side of the British as the wronged party. Naval vessels on both sides were sunk, and the conflict escalated dangerously, but it is difficult to see how United States forces could have become involved unless the British blockade of the Argentine coast had somehow involved the Soviet navy.

The Reagan administration's rhetoric on Central America has included threats of "going to the source," in Cuba, of support for insurgencies on the mainland, but it seems unlikely that Congress, with an ear to popular sentiment, would support any further involvement of the United States military beyond military supplies, training in the United States for indigenous forces, and a very limited number of advisors on the scene. All these demands can be met without additional forces beyond those required for the roles and missions already described.

Finally, this account would not be complete without some mention of the domestic functions of the military establishment. These functions fall into two unequal parts: one, the regular civil function of the Army Corps of Engineers; the other, the extraordinary function of maintaining law and order when local and state police cannot cope.

The Army Corps of Engineers has the responsibility for navigation and flood control of United States waterways. In pursuit of that responsibility, it has spent over $2 billion in each of the past two years. Waterway control is maintained separately from the military functions of the Corps and is highly decentralized. Because of its nature, it involves extensive contacts and negotiation with local interest groups, often pitting environmentalists against business and industrial lobbies. Recent studies indicate that the traditional dam-building approach of the Corps has been somewhat modified to subsume projects that involve relocating urban developments away from flood plains, rather than merely pouring concrete to control the effects of flooding.

The military has not been called upon to intervene in civil disturbances in recent years nearly as much as in the sixties. The years 1967 to 1970 were ones of particular domestic turbulence. Starting with the Detroit riots in the summer of 1967 and ending with the tragic Kent State incident in the spring of 1970, troops were committed on no fewer than five occasions during that period. In the Detroit riots alone, 4,300 National Guard troops and 655 members of the 82nd Airborne Division were committed.

There is an enormous range of situations in which the United States might conceivably want to deploy military power in pursuit of national interests; and in most of those situations it is highly unlikely that United States military force would ever be employed.

If military force were employed, there is a significant possibility that the crisis would escalate to a nuclear confrontation, from which neither side would emerge as victor.

That is a long way from making the world safe for democracy, or from fighting a war to end war. To the contrary, the United States military today is being asked to prepare to fight any and all wars, in order to avoid having actually to fight a war that might be the end of civilization.

III

Gauging the Establishment's Reach

THE MILITARY ESTABLISHMENT as an organization is often visualized as a monolith. In reality, it is more like a modern structure of prestressed concrete, held together by the tensions between opposing forces. From 1947, when it was first created under the name of the National Military Establishment, until 1961, the Department of Defense was a loose confederation of independent fiefdoms, uneasily presided over by the secretary of defense. The army, the navy, the Marine Corps, and the air force controlled their own budgets and determined their own force structures, with a minimum of coordination among them, although the president and the secretary of defense did determine total spending. Discussion had taken place on how the Department of Defense should be organized to deal with the problem of strategic nuclear war, and to develop new weapons systems, but the controversies that most intensely engaged their protagonists were the quasi-theological questions of roles and missions: Were missiles a new form of artillery, and therefore within the proper domain of the army, or a new kind of airplane and, therefore, the preserve of the air force? Could the army properly use close-support aircraft painted army colors and flown by pilots in army uniforms, or should these planes be painted air force colors and flown by pilots in air force uniforms? How far should the marines carry amphibious warfare beyond the sand and pebbles on the beachhead?

At the inception of the Defense Department, these controversies were exacerbated by the shrinkage of army manpower from a World War II high of 8,226,373 men in 1945 to a low of 591,487 in 1950, just before the Korean War. The most violent controversies arose over scarce congressional appropriations, as in the 1949 fight between the carrier admirals and the air force proponents of strategic bombing when the dispute was carried into the press and rival columnists were enlisted by the opposing sides.

These controversies extended to intraservice rivalry as well. Within the army, the "technical services"—such as the adjutant general, chemical, finance, ordnance, quartermaster, and transportation corps—had for decades successfully resisted efforts to streamline and integrate their operations. In the navy, the old-line bureaus presented the same problem. Even the newest service, the air force, suffered from the often successful efforts of the Strategic Air Command to dominate the entire organization.

Some small beginnings to resolve these conflicts were made in the 1949 amendments to the National Security Act, which established the legal basis for a single Department of Defense, and in the 1953 and 1958 Reorganization Acts, which strengthened the position and authority of the secretary of defense.* The seventh secretary, Thomas S. Gates, had taken a firmer hand than his predecessors in reducing rivalries. But not until 1961, under Secretary Robert S. McNamara, was essential new machinery created, particularly the Five-Year Defense Program. This program was an enormous intellectual and organizational effort that provided a functional breakdown of all Defense activities, organized by sector and sub-sector, combining elements of the different services and including, for each major weapon system, costs of operation, maintenance, and modernization, together with costs of development and procurement. Long lead time costs were, and still are, projected beyond five years, and the whole system was designed to be revised and updated at least annually to help resolve conflicts in a more orderly fashion.

*Discussed in more detail in chapter 6 below.

The systems analysis shop was created within the Office of the Secretary of Defense to provide the secretary with the analytical tools to evaluate proposals from the individual services and to take the initiative with new proposals. The services themselves were reorganized, and the army technical service and navy bureau structures substantially overhauled. The rationale of the new system was that if the civilian managers in the Pentagon were not relying on overall spending ceilings (and they claimed, on presidential authority, that they were not), they could at least exercise some more intelligent control over the separate pieces of the budget. That rationale may have run counter to political realism, even at the time. But the systems and procedures instituted by Secretary McNamara in the Kennedy administration have still survived, in perhaps less vigorous condition, to help civilian top managers get better control of inevitable bureaucratic tendencies to "suboptimize"—to emphasize perfection of the part over integration of the whole.

What has emerged is still by no means monolithic. As chart 3.1 makes clear, the line of operational command of forces in the field goes from the president and the secretary of defense through the Joint Chiefs of Staff to the commanders in chief of the eight field commands.* The military departments—army, navy, air force—train and supply personnel and develop and procure the equipment for the operating commands, but there are no units under the direct operational authority of any military department. While this is an accurate description of existing arrangements, it is not the whole truth. If chart 3.1 were shown to field-grade officers—colonels and majors—now on active duty, many of them would deny that it represented the formal command arrangements and would insist that the military services do control their own troops in the field. For these officers, the Department of Defense is not an overarching

---

*These include five unified commands (including elements of several services)—the Atlantic, European, Southern, Pacific, and Readiness (which provides reinforcement and augmentation in response to worldwide contingencies)—and three specified (or single-service) commands—the Aero Space Defense, Strategic Air, and Military Airlift. The Rapid Deployment Joint Task Force will become a sixth unified command on January 1, 1983.

organization of which their service is a part. When they speak collo-quially of the department, or "DOD," they do not customarily in-clude their own organizations, but only the secretary of defense and his staff (technically the Office of the Secretary of Defense, or OSD), a separate and alien entity, rather like a foreign power with which their organization is forced occasionally to deal.

Field-grade officers' interpretations of the lines of command are partly correct. The people who do work for the Joint Chiefs are the roughly 1,300 individuals, including some 700 military officers or their civilian equivalents, of the Joint Staff. These officers pre-pare the memoranda on proposed policy issues, take them through the various stages of evolution—first the buff, then the flimsy, the green (usually through several drafts), the purple (only when there is a major objection), and finally the red stripe memoranda that go to the secretary of defense. But the Joint Staff doesn't do the real work. That is done by the staff in each military department who prepare the service position on every issue, and these positions are then negotiated, if possible, within the Joint Staff. Thus, despite the size of the Joint Staff, each service may have as many as 30 (or even more) members of its own staff dedicated to full-time JCS support. The primary role of each member of the Joint Chiefs, except the chairman, is to serve as chief of his own service and he is briefed for the weekly chiefs' session, not by the Joint Staff, but by his own service staff.

This arrangement is what Congress had in mind when it amended the military unification legislation in 1958 to include a specific prohibition against the creation of a single general staff, and the numerical restrictions on the size of the Joint Staff are intended to carry out this explicit policy. Congress feared the power of a fully militarized Department of Defense and sought to perpetuate in-terservice competition, or, as it has been put more cynically, a situation in which one service could be played off against the others. This competition may be an important tool in preserving civilian control, as discussed in chapter 6 below.

On the other hand, such competition also may breed a sort of decision-making inertia in which consensus on specific issues is

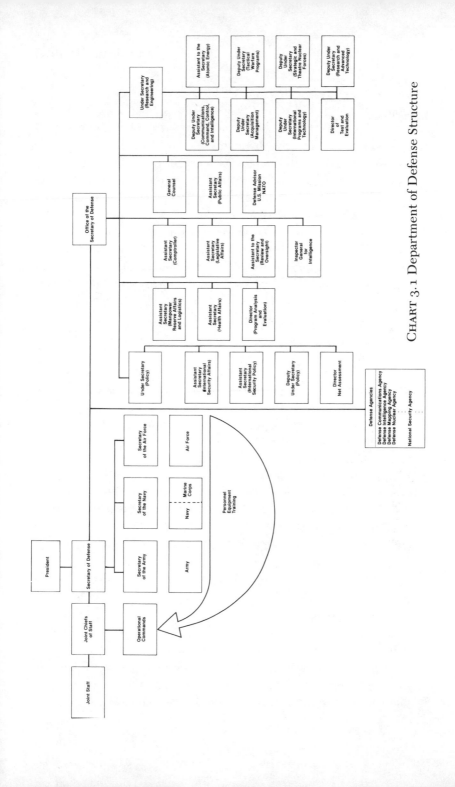

CHART 3.1 Department of Defense Structure

rarely achieved and compromise therefore becomes the sine qua non for bureaucratic survival. This condition has prompted former Joint Chief of Staff Chairman David C. Jones and others to argue for a major overhaul of the joint staff system; these proposals are addressed in more detail in chapter 7.

In time of extreme budgetary uncertainty, interservice competition may produce particularly wasteful redundancy. Each of the services maintains its own logistics and training systems. The army and the navy run separate helicopter-training schools, costing the Pentagon $63 million a year more than the expense of a combined school. There are four separate air forces to maintain—one for each service. Excess repair depot costs are estimated at $250–$400 million annually. Nonstandardized ground repair equipment for aircraft maintenance probably produces excess costs of $300 million a year. The insistence on independent supply functions by each service costs about $100 million each year and produces as many as 4,000 "superfluous" jobs.

In the competitive process, the services provide a powerful assist. In 1961, for example, the secretary of defense transferred the strategic intelligence staffs out of the military departments to form the Defense Intelligence Agency (DIA), reporting to the secretary of defense through the Joint Chiefs. This was done on the ground that strategic intelligence was a function of the operating commands and should be placed in operations, rather than with training and supply. DIA was created to coordinate and consolidate the individual views of the service intelligence staffs and eventually to replace the staffs in toto; this never happened. When the intelligence staffs were transferred out of the military departments, the departments began to build up their staffs in areas called euphemistically "foreign technology," so that they were able to compete, to some extent, with the new interservice agency. Today, DIA does little information collection of its own (aside from operating the overt system of military attachés overseas) but instead processes raw data from the services into reports circulated within the Pentagon and the rest of the intelligence community. DIA employs approximately 5,000 persons, while the air force intelligence service employs about 10 times as many and outspends DIA roughly 25 to 1.

By the same token, the legislative liaison staffs of the individual services vastly outnumber the legislative liaison staff of the secretary of defense—and the secretary's efforts to reduce their numbers have thus far been unavailing. Tampering with these staffs means intervening directly in long-established relationships with key legislators and congressional committees. In an organization the size of a military department, the organization itself can compensate for externally imposed reductions. Service links with Congress, and particularly with the Armed Services Committees and their senior members and staff, are closer than the links between the secretary of defense and his staff and Congress and congressional staff. The controversy over these relationships resulted in the disbanding in 1965 of the Capitol Hill active reserve units, including the famous 999th Air Force Reserve Squadron commanded by Major General (Senator) Barry M. Goldwater, USAF, which permitted eighty-three congressmen and senators to spend short periods of active duty in such prime military observation posts as London and Paris. But the 999th was only a surface manifestation of a more deep-seated and persistent phenomenon.

In the field, the joint operational commands exercise more authority in form than they do in fact over their separate service components. Until recently, the principal staff officers of the service components of a joint command—for example, the deputy chief of staff for operations, United States Army, Europe—would outrank their opposite numbers for the entire joint command itself. Even in Vietnam, the complaint was frequently made that service interests were not subordinated to common concerns. Military observers complained bitterly that the Pentagon practiced business as usual; that the military departments did not give the war priority over the internal needs of the military organization; that command rotation policies, for example, were designed to give everyone a turn, rather than to achieve the most efficient level of operations; and that departmental research and development programs were not sufficiently oriented to the needs of the war.

During the post-Vietnam era, it has become increasingly apparent that both inter- and intraservice differences are fragmenting even further the focus of the military establishment. This emerging reality is modifying the character of military policy and hardware decisions. What have evolved are floating coalitions that cut across the services, government entities, private corporations, and even foreign civilian and military structures. The decision-making process is rarely confined to the Department of Defense.

Beyond the Department of Defense, the organizational picture is even less monolithic. The Pentagon is only one part of a system of often opposed governmental units dealing with the problem of national security. Within the executive branch proper, these elements include the Department of State, the National Security Council staff, and the White House. Mention must also be made of the Office of Management and Budget, the Central Intelligence Agency, the Department of Energy, and the Arms Control and Disarmament Agency.*

At the other end of Pennsylvania Avenue, there is Congress, which includes the 535 senators and representatives, the greatly expanded congressional staff, and the special legislative branch agencies, among them the long-established General Accounting Office, the Congressional Reference Service of the Library of Congress, and the six-year-old Congressional Budget Office.

And reaching into Washington are the nongovernmental elements of the military establishment: the various military alumni associations; the industrial part of the military-industrial complex immortalized in President Eisenhower's farewell remarks; and the academic and quasi-academic research organizations that are more-or-less dependent on military largesse for their continued existence.

The extensive involvement of civilian authorities and agencies in the management of the military establishment contrasts sharply with the situation in the Soviet Union, where the military is much more compartmentalized away from the rest of the government. In the course of one arms control negotiation in Geneva between the United States and the Soviet Union, Soviet military representatives reportedly cautioned their American counterparts against discuss-

---

*As the potential for warfare in space grows, the National Aeronautics and Space Administration, though it presently cannot claim official acknowledgement, is also likely to become an unofficial extension of the defense establishment.

ing certain matters with the civilians on the Soviet negotiating team. American scholars have often remarked on the difficulty of finding Soviet civilians who could engage in informal discussion of military strategy.

In theory, the Department of State should be involved in every military decision that has significant impact on the foreign relations of the United States. It is difficult to imagine a military decision of any importance that would not have such an impact. When Dean Acheson and George Marshall were serving as secretary of state and secretary of defense respectively, they made a pact that neither one would ever say to the other, "This is a purely military matter," or "This is a purely political matter." In the absence of this meeting of the minds between the two secretaries, however, the State Department's involvement is limited in at least two ways:

Decisions that can have profound effects on foreign relations may be taken in a context where there is no obvious occasion for consultation with the State Department (e.g., a decision on the specifications for a new weapon system, or a change in the force structure). If a decision involves ends rather than means, consultation will extend beyond the Pentagon, but too often ends are affected, and perhaps even determined, by the choice of means. Where consultation with allies or negotiation with adversaries is involved, the State Department must be, and almost always is, brought into the act. But on even more critical decisions, as in the initial stages of weapons developments when the fundamental direction of a new weapons system is being determined, the State Department may not even be aware of what is going on. For example, Secretary of State Muskie was not made aware of the deliberations leading up to the promulgation of Presidential Directive 59, in the summer of 1980, on the United States targeting policy for nuclear weapons.

Second, the State Department is not well equipped for competition with the Department of Defense in the bureaucratic arena. Military men are primarily management-oriented, while the main function of foreign service officers is to observe and to report. The State Department has been described as suffering from a severe case of heliotropism: the secretary of state sees his primary function as advising the president, not as leading the department. The under secretaries and the assistant secretaries see their primary function as advising the secretary, not as leading their bureaus; the same applies to the office heads and section chiefs. There is no one who sees his first responsibility as running the store. The reverse is true of the Defense Department, where the primary emphasis is on the leadership and on achieving goals prescribed by higher authority. In interagency combat, the results are likely to favor the better managers. The imbalance is further skewed by the larger resources available to Department of Defense out of a budget almost 100 times greater. There is inevitably more leeway for the collection of all relevant information in a hurry, for quantifying what is difficult to quantify, and for getting it down on paper smoothly, accurately, and convincingly.

The one part of the State Department that is specifically devoted to relations with the Pentagon—the Bureau of Politico-Military Affairs—has been sufficiently bruised by its encounters with a stronger force that too often it serves as an apologist for the Pentagon position within the department. Instead of marching to the beat of a different drummer, the bureau is likely to follow the military band. And because the secretary of state has so much more to do than the secretary of defense yet has no principal civilian managers quite like the secretaries of the military departments to whom he can delegate, he has, vis-à-vis the defense secretary, a competitive disadvantage for which his seniority in the Cabinet does not compensate.

The second major actor in the interdepartmental arena is the National Security Council staff. The staff must be distinguished from the National Security Council itself, which, from its creation in the National Security Act of 1947, has been a stage for political display rather than an arena for actual decision making. But the staff, which numbered some 72 permanent members under Zbigniew Brzezinski in the early days of the Carter administration (including a press officer and a congressional liaison), remains today an independent force, even at a reduced size of 61 budgeted positions.

The experimental interlude during which Brzezinski's successor, Richard Allen, was forced to maintain a low profile, ended after only a year, when Allen was put out to pasture. He was replaced by Deputy Secretary of State William Clark, who was given the national security advisor's job complete with expanded duties and renewed direct access to the president.

In theory, the assistant to the president for national security affairs and his staff are only a neutral communications link between the departments and agencies involved in the national security process and the president, transmitting their recommendations, summarizing their disagreements, and, where appropriate, transmitting back the president's instructions. In actual practice, the staff often acts as a minidepartment of its own, participating actively in interdepartmental working groups, and pressing its own views on policy issues. Although this role may be resented by the other agencies, there is not a great deal they can do about it. Strong cabinet officers in Defense and State, able and willing to delegate a good deal of their day-to-day management responsibilities to subordinates and preferably with political constituencies of their own (as was the case with Secretary of State Muskie), can insist on their prerogatives. But in the last analysis it is the president who determines the scope of authority of his national security advisor.

The president is of course the final arbiter of all important decisions, and, although much depends on how questions are shaped before they reach the president, he has the power not only to select the answer but to change the shape of the question. As noted recently by Amos Jordan of the Georgetown University Center for Strategic and International Studies: "Today, the President needs a National Security Council staff that is capable of defining policy alternatives in ways that reflect his broad interests and not those of a particular government department or agency. Those formulations and options need to be presented to the President in a manner that includes his political interests and the interaction of domestic and foreign policy, and that reaches beyond the responsibilities of the Secretary of State."

Secretary McNamara came to President Kennedy in the early

sixties with a decision he had hammered out with the Joint Chiefs of Staff on the number of Minuteman missiles to be deployed. He reported that, after a good deal of debate, he had persuaded the chiefs to accept a figure of 1,000 rather than 1,200. President Kennedy asked the secretary, "Why not 500?" The point is not that the president proposed a significantly lower figure—and indeed in the end he accepted the figure 1,000—but that he was able to question the premises of the debate.

On the other hand, a president who waits too long to intervene in a policy dispute may find some decisions foreclosed or may have to accept political penalties that would not otherwise have been imposed. When President Carter rejected the recommendation brought to him by his national security advisor, after extensive consultation throughout the government, for the production and deployment of the so-called neutron bomb, he provoked a wave of criticism. The reaction was in response not to the merits of his decision but more to the fact that a contrary decision had been anticipated in the domestic and international political processes, and it was late in the day to try to develop a constituency to support the president, no matter how sensible his reaction when he eventually faced up to the proposal.

The Office of Management and Budget has a key role in the decision-making process since it represents the president in reconciling the competing demands of the various departments for their shares of the overall federal budget. An aggrieved department head can always appeal from an OMB decision to the President himself, but each appeal uses up a limited amount of political capital, and even the secretary of defense will think twice before challenging the president's budget director. Where the budget director's authority may be successfully challenged by the secretary of defense is on the broadest questions of total defense spending, as was demonstrated when Secretary Weinberger was able to defeat Budget Director David Stockman, even before Stockman's fall from grace.

Where the Office of Management and Budget stands guard over the flow of dollars into the military establishment, the Central Intelligence Agency stands guard over the flow of information and judgments about the state of the world. The Defense Department does not rely exclusively or even primarily on the CIA for facts and opinions about other nations' military forces, the prospects for peace or war, or about the condition of other political and economic systems. The department has its own elaborate intelligence apparatus, including the Defense Intelligence Agency, described above, and the highly secret National Security Agency, which employs complex technology to listen in on friends and potential foes around the world. But the CIA is the predominant source of foreign intelligence and often competes with and differs in judgments from the DIA, even on technical subjects. Further, the head of the CIA, wearing his hat as director of central intelligence (DCI), chairs the National Foreign Intelligence Board (NFIB), the senior management group that formulates the official positions of the United States intelligence community on the capabilities and vulnerabilities of our allies and potential enemies—although the military members of NFIB may, and frequently do, dissent from community judgments. The effective exercise of this right to dissent assumes, of course, that the NFIB is permitted by the DCI to function as frequently and as broadly as it might, something which some DCI's reportedly have not done.

Even more important, the director of central intelligence sits in on the meetings where recommendations on national security policy are put in final form for presidential decision, and his subordinates participate regularly in lower level meetings. While the director may choose to abstain as the other participants take sides in policy debates, his views on the underlying intelligence assumptions behind the decisions are taken very seriously. On questions such as the verifiability of arms control agreements, the director's views are critical for the political viability of such agreements in Congress and in the nation.

One of the special aspects of the intelligence community is its authority to withhold details of information on national security grounds. Abuse of the "national security" label has been quite effectively discredited since Watergate, and legitimate use of the privilege carries little political advantage. But the other side of the coin—the selective disclosure of classified information overtly or through leaks—is a basic weapon in political infighting, both within the national security establishment and with other departments and branches of government. The Carter administration's decision, during the 1980 election campaign, to announce the development of the Stealth bomber on the ground that information about the bomber had already been leaked, was a new twist in tactics. Similarly, the Carter administration, during the 1980 election campaign, created a new intelligence category, aptly (or ineptly) called "Royal," in order to protect "politically sensitive information" regardless of its source—apparently an attempt to discourage leaks from within the government by enemies of the administration. Such an attempt is not likely to succeed, since recent history suggests that most leaks come from the top or at least from so high a level that compartment walls cannot prevent them. When Henry Kissinger's penchant for supersecrecy excluded the Joint Chiefs of Staff from information to which they felt they were entitled, it was reported that they infiltrated a spy as a clerk at the National Security Council and thus kept up to date in spite of Kissinger.

Charges have been made of skewed or even distorted reporting by the CIA in order to tell the president what he supposedly wants to hear, or simply to minimize conflict with the Defense Department or with other elements of the national security community. An unhappy case in point was the apparent overruling of CIA staff by senior agency officials when their figures did not jibe with lower Pentagon estimates of Viet Cong strength. But, to the extent that one can guess at the classified facts, it is more often the case that unpleasant advice from the CIA is overlooked and the political leadership allows itself to be seduced by the professional "can do" approach of the military.

In issuing his two recent executive orders on December 4, 1981, one to govern the activities of our intelligence agencies, the other to reestablish the Intelligence Oversight Board (which is charged with ensuring the legality of United States intelligence activities), President Reagan observed:

It is not enough . . . simply to collect information. Thoughtful analysis is vital to sound decision-making. The goal of our intelligence analysts can be nothing short of the truth, even when the truth is unpleasant or unpopular. . . . When there is disagreement, as there often is, on the difficult questions of our time, I expect those honest differences of view to be fully expressed.

To the extent that the president adheres to his commitment, and the intelligence community conforms to his direction, this will represent a signal achievement.

The Department of Energy (DOE) performs a specific function within the military establishment. It is responsible for the development and production of nuclear weapons as well as the production and control of the weapons-grade fissionable materials that go into those weapons. The two great nuclear weapons laboratories at Livermore and Los Alamos are operated under the sponsorship of the DOE, as successor to the original Atomic Energy Commission. The Energy Department's views on how many nuclear weapons we need, on how much nuclear material we should have in the pipeline and in the stockpile for future weapons tend to equal if not exceed the estimates of the Department of Defense itself. At present, the nuclear weapons program is roughly half of DOE's budget. If President Reagan has his way, DOE will fold its tents during the latter half of 1982 and transfer its nuclear functions to the Department of Commerce. Because such a move would place these sensitive functions under an agency without previous expertise in the area, resistance to the idea has arisen in Congress. Some senators have even gone so far as to suggest that the nuclear program should be run in laboratories under separate civilian authority.

In 1978–1979, during negotiations for the Comprehensive Test Ban Treaty between the United States, the United Kingdom, and the Soviet Union, controversy arose over the duration of the agreement and over the lower limits on the size of prohibited detonations. It was the Department of Energy that led the fight for a three-year rather than a five-year treaty and for a limit high enough to permit a considerable range of low-yield weapons tests. During the debate within the United States government over the United States position in the 1978 Special Session on Disarmament of the United Nations General Assembly, again it was the Department of Energy which most vigorously opposed putting forward a United States offer to attempt to negotiate a treaty terminating the production of weapons-grade fissionable materials—although official United States policy had favored such an agreement since the Eisenhower administration.

Finally, before we leave the executive branch, mention must be made of the role played by the Arms Control and Disarmament Agency (ACDA). ACDA was established by Congress in 1961 "to develop acceptable political and technical alternatives to the present arms race." The director of ACDA is the principal advisor on arms control and disarmament matters to the secretary of state, the National Security Council, and the president.

ACDA serves a number of important functions. Consistent with the secretary of state's responsibility for the conduct of foreign affairs, the agency has the authority for the management of international arms control and disarmament negotiations; it leads United States delegations to most such negotiations and it plays a leading role in preparing and coordinating United States negotiating positions.

Organizationally, ACDA is divided into four major bureaus: (1) the Bureau of International Security Programs, which handles the diplomatic, political, and technical aspects of major strategic bilateral and regional arms control negotiations, such as SALT; (2) the Bureau of Multilateral Affairs, which handles arms control negotiations that are or will be negotiated in multilateral forums, such as the Committee on Disarmament and the United Nations; (3) the Bureau of Non-Proliferation, which contributes to the government's efforts to curb the spread of nuclear weapons; and (4) the Bureau of Weapons Evaluation and Control, which assesses the policy implications and arms control impacts of United States military technology programs and their transfer overseas.

Aside from the salutary accomplishments of individual arms control negotiators, ACDA has not exerted a significant impact on the defense establishment in its 20 years of existence. Lacking the bureaucratic clout, technical and intellectual depth, and access to data necessary to deal on equal terms with Pentagon or even with its opposite numbers in State and the NSC, the agency's role has been, with lonely exceptions, an ineffectual one.

One of ACDA's more time-consuming activities is the preparation of so-called Arms Control Impact Statements on major weapon systems being developed by the Pentagon. The theory of these statements, which are transmitted to Congress in classified form and published in "sanitized" or unclassified form for the general public, is that they will provide a check on weapons developments that might not otherwise take sufficient account of the arms control problems they were creating. In practice, however, the process becomes an unequal tug of war between ACDA and the other parts of the national security establishment within the executive branch which have decided to proceed with a weapons development program and don't want ACDA putting a monkey wrench in the process of congressional approval. The claimed analogy between the Arms Control Impact Statement and the Environmental Impact Statement is not a valid one since the EIS is prepared by the entity proposing a particular course of action and is by and large a useful exercise in public justification. It has been suggested that ACDA could well afford to give up preparation of Arms Control Impact Statements in exchange for a regular seat at the table, probably somewhere in the bowels of the Pentagon, where initial decisions are actually made about the size and shape of new weapon systems.

In the last analysis, the Department of Defense looks to Congress for the flow of dollars that it needs to sustain its operations. To be sure, there are wide fluctuations over time in congressional support. These fluctuations are in part a function of administration policy but they are also a function of congressional attitudes. The climate at both ends of Pennsylvania Avenue is significantly affected by events at home and abroad: the temperature of the Cold War and the vigor of the domestic economy. Spending more on defense is a simple way to respond to popular concern among the electorate about national security, although the specific objects of expenditure may or may not reassure our allies or deter our antagonists. And where the executive branch can rattle the sword, the legislative branch can only shake the purse at foreign adversaries.

The military has been as sensitive to the moods of Congress as have the other parts of the executive branch bureaucracy, and, at least since the 1950s, it has had more resources to maintain congressional liaison officers who are always available to help with constituent problems or even with congressmen's own needs. But three factors have in recent years made the symbiotic relationship between Congress and the military more an arms-length one, while at the same time amplifying the role of Congress in the business of government. Two factors are procedural, growing out of changes in the way Congress staffs itself and organizes to handle the budget, and the third is substantive, growing out of the economic problems that attend the budget process.

The number of congressional committee staffers has more than doubled in the eleven years since 1971. The effects of this increase are physically demonstrated in the construction and leasing of additional office space, while staff members are crowded into smaller and smaller cubicles. The political effects are demonstrated in an increasingly intimate involvement of congressional staff in the details of executive branch management, to an extent that has been sometimes described as "micro-management." Congressional staff members will involve themselves in the details of development and procurement problems which their bosses are too busy to follow; the relationships that result, while they may have constructive consequences, are more likely to be antagonistic than cooperative.

The arms-length relationship between the executive branch bureaucracy and the staffs of individual congressmen and congressional committees is reinforced by the increased activity and professionalization of the General Accounting Office, particularly under the guidance of Elmer Staats, who served as comptroller general from 1966 to 1981. The critical, analytical role of the GAO vis-à-vis the Pentagon is not unlike the sixties role of Secretary McNamara's program analysis staff vis-à-vis the military services. This role should become increasingly typical for the GAO as the old-line accountant types who have been with GAO since its inception give way to a new brand of more technically-oriented analysts.

The creation of the House and Senate budget committees in 1974, as part of the reorganization of the legislative budgetary process, has required Congress, by its own act, to face up to the budgetary impact of specific increments in defense spending. The budget committee recommendations, when acted on by both houses in the joint budget resolution, force Congress to accept a self-imposed ceiling on the military budget that can only be raised by a corresponding reduction somewhere in the nonmilitary budget or by a concomitant increase either in revenues or in the budget deficit. The public and objective character of this part of the budget process tends to reduce the effectiveness of private lobbying by one or another of the military services (or elements within one of the services), since an increase in one area has to be matched by a cutback in another unless the budget resolution itself is amended.

At the same time, inflationary pressures and political resistance to increased government spending have intensified the budget crunch. Military spending is still the largest part of the overall budget that is not determined in advance (unlike so much of the budget of the Department of Health and Human Services). Traditionally, agricultural subsidies have been reduced in order to avoid the need to balance the budget by cutting defense spending. This expansion joint in the budget has now been diminished since agricultural subsidies no longer bulk so large. In the fifties, a Democratic Congress pressed a reluctant Republican administration for more defense spending. In the eighties, a predominantly Republican Congress seems to have growing concern about a Republican administration's bigger defense spending plans.

This is not to say that Congress no longer mandates specific increases in the Defense Department budget over and above already increased administration recommendations. But dramatic confrontations like the Rose Garden meeting between President Kennedy and Chairman Vinson of the House Armed Services Committee, when the two almost came to an impasse on additions to the fleet, are probably a thing of the past—even should a less defense-minded administration succeed to office.

Like every other entity in the public sector, the Pentagon has its private sector supporters, beginning with the major veterans organizations: the American Legion and the Veterans of Foreign Wars; the alumni associations of the four services—the Association of the United States Army, the Navy League, the Marine Corps League, the Air Force Association; and the National Guard Association and the Reserve Officers Association. These organizations subsist on dues from individual and corporate members and from defense contractors who advertise in their publications. The associations support the legislative interests of the services with which they are affiliated by employing large professional staffs. These employees work closely with the lobbyists for major companies engaged in the defense industry. Industry representatives in turn seek to earn congressional—as well as executive branch—support for the weapons developments in which their companies have invested time and money. Lobbying expenses for the defense industry are generally reimbursable and are thus absorbed by the taxpayers, although Secretary of Defense Weinberger issued a 1981 directive attempting to distinguish between "lobbying"—an unallowable expense—and "legislative liaison"—an allowable expense. The *Washington Post* staff writer reporting on the directive questioned whether it would eliminate practices like that of Rockwell International, which "charge[d] DOD for a study of the economic impact of the B-1 bomber—broken down by congressional districts."

In addition to individual company lobbyists, the defense industry is represented by such organizations as the American Defense Preparedness Association, the National Security Industrial Association, and the Aerospace Industries Association. Finally, representatives of laboratories and research organizations, profit and nonprofit, press for further funding to support the lines of inquiry on which they have been working.

The catalog of entities affecting the military budget and the force structure would not be complete without some reference to the impact of alliance politics. Some of the most bitter battles within NATO are fought over the choice of weapon systems and their components (and the consequent selection of one manufacturer over another) both among the member countries and within those countries. The interaction among national, regional, service, and company interests is increasingly a feature of the weapons development process. To help alleviate such problems, a number of initiatives have been undertaken by NATO. These include *memoranda of understanding*, designed to eliminate "buy national" restrictions and to open markets to reciprocal competition; the *families of weapons concept*, which involves dividing responsibilities for developing particular types of weapons among a number of nations; and *dual production*, by which a nation that has developed a system useful to other alliance nations would permit the others to produce the entire system or portions thereof. The ultimate success of such initiatives remains to be seen; although they represent attempts to ameliorate *inter*national divisiveness, these measures can have devastating effects on *intra*national policies and decisions.

What should be clear from the foregoing discussion is that, as the nation's largest institution, the military establishment not only reflects but also amplifies the most persistent trend in all public decision making: the growth of an elaborate, ritualistic process, the net effect of which is to fudge accountability and to make speedy and clear decisions extremely difficult. Each element is administered by people who too often lack the necessary sense of responsibility for making the whole enterprise work. The military establishment, then, is no simple pyramid of power. But its base is more than broad enough to accommodate a good deal of internal conflict without diminishing its overall impact on American society.

IV

The Impacts on the Economy

THE MILITARY ESTABLISHMENT as a segment of the economy can be described with considerable precision. As table 4.1 shows, expenditures by the Department of Defense since 1961 in constant dollars and as a percentage of the gross national product have been substantial and projected increases through 1984 are even more substantial. The 1970s saw three distinguishable periods of military spending: a rapid decline in real spending during the late Vietnam era (1968-1973); a leveling off in the post-Vietnam era (1974-1978); and a period of renewed growth in the current rearmament era, projected to reach even higher levels by the mid-eighties. At the peak of the Vietnam era, the military absorbed 9.5 percent of the gross national product, but by 1980 military outlays accounted for only 5 percent of a (much larger) GNP. The Reagan administration projects 1984 military spending at $341 billion, or 7.1 percent of the GNP, and many observers believe this estimate does not take sufficient account of variable cost overruns or of the even greater impact of inflation on military expenditures than on the general economy.

To the extent that sources of funding for the military establishment extend beyond the Department of Defense, they are included in table 4.2.

The Kennedy and Johnson administrations went to considerable lengths to distinguish between the military space program

TABLE 4.1 Department of Defense Expenditures and GNP, Fiscal 1961–1982 (*Billions of Dollars*)*

Fiscal Year	GNP	U.S. Gov't Total Outlays	DOD Actual Expend. (Outlays)	As % of GNP	As % of Total Gov't Outlays	Total Gov't Outlays (1972 Dollars)	DOD Outlays (1972 Dollars)	As % of Total Gov't Outlays (Real)
1961	523.3	97.8	46.6	8.9	47.6	157.1	74.8	47.6
1962	563.8	106.8	49.0	8.7	45.9	168.5	77.2	45.8
1963	594.7	111.3	50.1	8.4	45.0	170.0	76.8	45.2
1964	635.7	118.6	51.5	8.1	43.4	176.6	77.0	43.6
1965	691.1	118.4	47.5	6.9	40.1	173.0	69.3	40.1
1966	753.0	134.7	54.9	7.3	40.8	188.1	76.6	40.7
1967	796.3	158.3	68.2	8.6	43.1	212.3	92.3	43.5
1968	868.5	178.8	78.8	9.1	44.1	229.8	101.7	44.3
1969	935.5	184.5	79.4	8.5	43.0	223.3	98.1	43.9
1970	992.7	196.6	78.6	7.9	40.0	220.6	90.3	40.9
1971	1077.6	211.4	75.8	7.0	35.9	223.3	81.5	36.5
1972	1185.9	232.0	76.6	6.5	33.0	232.0	76.6	33.0
1973	1326.4	247.1	74.5	5.6	30.1	233.3	69.9	30.0
1974	1434.2	269.6	77.8	5.4	28.9	232.8	68.1	29.3
1975	1549.2	326.2	85.6	5.5	26.2	255.4	68.4	26.8
1976	1718.0	366.4	89.4	5.2	24.4	268.4	66.9	24.9
1977	1918.0	402.7	97.5	5.1	24.2	273.8	67.2	24.5
1978	2156.1	450.8	105.2	4.9	23.3	284.4	67.5	23.7
1979	2413.9	493.7	117.7	4.9	23.8	282.5	70.0	24.8
1980	2626.1	579.6	135.9	5.2	23.4	294.9	72.5	24.6
1981	2922.0	655.2	159.8	5.5	24.4	296.7	77.8	26.2
1982	3160.0	695.3	187.5	5.9	27.0	288.4	83.0	28.8

*Columns 2–5 are current dollars
1982 figures are estimated

TABLE 4.2 National Security Expenditures,
Fiscal 1961–1982 *(Billions of Dollars)*

Fiscal Year	DOD Outlay	Atomic Energy Defense Activities[3]	Veterans Admin. Outlay	Security Assistance Outlay[2]	Selective Service Outlay	Total
1961	46.6	(1.3)	5.4	(1.4)[3]	.03	52.0
1962	49.0	(1.4)	5.4	(1.4)[3]	.04	54.4
1963	50.1	(1.3)	5.2	(1.7)[3]	.03	55.3
1964	51.5	(1.3)	5.5	(1.5)[3]	.04	57.0
1965	47.5	(1.6)	5.5	(1.2)[3]	.04	53.0
1966	54.9	(1.2)	5.1	(1.0)[3]	.05	60.1
1967	68.2	(1.1)	6.1	(0.9)[3]	.06	74.4
1968	78.8	(1.2)	6.6	(0.6)[3]	.06	85.5
1969	79.4	(1.4)	7.7	(0.7)[3]	.07	87.2
1970	78.6	(1.4)	8.7	(1.1)[3,4]	.08	87.4
1971	75.8	(1.4)	9.6	(1.5)[3,4]	.08	85.5
1972	76.6	(1.4)	10.7	(1.4)[3,4]	.08	87.4
1973	74.5	(1.4)	12.0	(1.5)[3,4]	.08	86.6
1974	77.8	(1.5)	13.2	1.7[4]	.05	92.8
1975	85.6	(1.5)	16.6	2.3[4]	.05	104.6
1976	89.4	(1.6)	18.4	1.7[4]	.04	109.5
1977	97.5	(1.9)	18.0	1.6[4]	.01	117.1
1978	105.2	(2.1)	19.0	2.4[4]	.01	126.6
1979	117.7	(2.5)	19.6	2.3[5]	.01	139.6
1980	135.9	(2.8)	21.1	2.8[5]	.01	159.8
1981	159.8	(3.4)	22.9	3.1[5]	.03	185.8
1982[1]	187.5	(4.5)	24.1	3.5[5]	.02	215.1

[1] 1982 figures are estimates.
[2] Does not include trust fund figures.
[3] Included in DOD outlays.
[4] Includes security supporting assistance (Agency for International Development).
[5] Includes economic support fund and peacekeeping operations outlays.

of the Department of Defense and the program of the National Aeronautics and Space Administration (NASA), but many major industrial entities that do business with NASA also depend on the Defense Department as their only other important customer in the aerospace industry. And since NASA pays its share of fixed costs as well as helping to balance out the peaks and valleys in defense production, there is some justification for including some portion of the NASA budget as part of defense costs. Because of the uncertainty of the calculation, however, NASA is excluded from the defense budget. Similarly, there is no item for the Central Intelligence Agency because its overall budget and the proportion devoted to clearly military purposes are both secret.

The people who work in the military establishment themselves constitute a market for food, clothing, and shelter; for automobiles and television sets; and, later, for retirement condominiums. As a group, their needs may be indistinguishable from those of the rest of the population, but in particular communities and in some large areas—e.g., southern California, tidewater Virginia—military establishment personnel dominate the economic landscape because of their sheer numbers. This dominance is most vividly apparent when a military base is abandoned or a big defense plant loses contracts. The comparative importance of the military establishment in each state is shown in maps 4.1 and 4.2, which illustrate defense contract awards and the ratio of defense payroll to personal income. Organizations that figure in the list of defense contractors range from General Dynamics, General Electric, and General Motors to small machine shops. To a greater or lesser extent, each is part of the military establishment.

For most of the people who receive defense dollars, those dollars are their primary connection with the military establishment. But for some of the most important members of the establishment—civilian political appointees, key senior officers—the cash nexus is their least significant connection. Some of the people who receive significant amounts of defense dollars—stockholders in businesses with mixed defense and commercial markets, academic administrators whose budgets include substantial defense contracts—may not think of themselves as part of the military establishment at all.

MAP 4.1 Defense Contract Awards by State, 1980

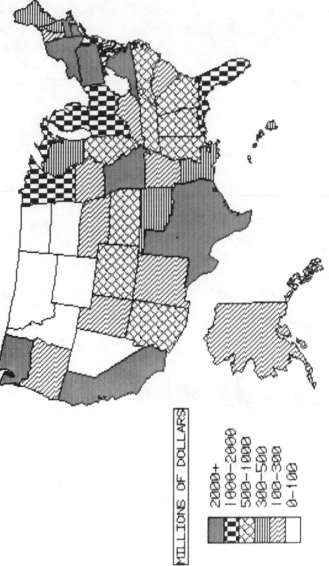

MILLIONS OF DOLLARS

2000+
1000–2000
500–1000
300–500
100–300
0–100

SOURCE: U.S. Dept of Commerce Statistical Abstract of the United States, 1981
Computer Graphics by Abt Associates, Inc.

MAP 4.2 Ratio of Defense Payroll to Personal Income by State, 1980

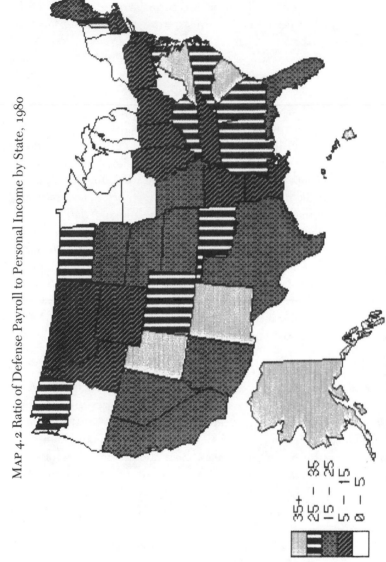

35+
25 – 35
15 – 25
5 – 15
0 – 5

SOURCE: U.S. Dept of Commerce Statistical Abstract of the United States, 1981 Computer Graphics by Abt Associates, Inc.

The major population elements in the military establishment are military personnel, direct-hire civilians, and employees of defense contractors. There is a fourth large category of persons primarily dependent on members of the first three groups: spouses and dependent children; tradesmen, from the military tailor to the lunchroom proprietor around the corner from the defense plant whose customers are members of the military establishment; the professionals—lawyers and accountants, bankers and brokers, public relations counselors, and trade journalists—who serve defense contractors; the retired military for whom the establishment provides a second income, and often the basis for their postmilitary retirement occupation; and others removed by various degrees.

A former Pentagon reporter for the *New York Times*, Jack Raymond, noted that the reach of the military budget is larger than is generally imagined. As he observed, "It provides, for example, $6,000 for flowers for American battle monuments." Flower growers, too, can be a part of the military-industrial complex. How far should such a list extend? Should it include: labor leaders of defense plant unions? Editors and publishers of newspapers in communities with substantial defense plants or military bases? Ministers of the principal churches in such communities? Members of and staffs of the armed services committees and military appropriations subcommittees of the House and Senate? Congressmen from districts with substantial defense business? Presidents of universities seeking to retain (or to attract) faculty whose research interests are to some extent dependent on defense research assistance? While it seems possible to define and measure the population of the military establishment, the fact is that the count of persons ultimately connected with the military is almost infinitely extensible.

Nevertheless, several conclusions can be drawn about the population of the military establishment. The drop from a peak of nearly 5 million in 1968 to 3.3 million in 1977 was entirely due to a reduction in military personnel. Civilian employment remained relatively stable through the post-Vietnam period, which saw major cuts in defense spending. At the same time, the proportion of the total defense budget for items other than pay and allowances increased significantly.

It was not always so that the military establishment bulked so large in American society and in the American economy. For the first century and a half of the Republic's history, the peacetime military was underfunded and unappreciated. Apart from the massive convulsion of the Civil War, the Republic's wars were brief adventures that left no lasting marks on the structure or composition of the military. Even World War I involved only 4.3 million men and a military budget at its peak of $11 billion.

The pre-World War II military amounted to no more than a small backwater in American society, out of touch with the rest of the country, often stationed in remote outposts in what was once Indian territory. Its men came to town on Saturday nights only to find signs in tavern windows telling them that their patronage was not welcomed. Officers kept to themselves in a world where pay was low and promotions agonizingly slow but one could get by with little effort or imagination. The wonder was that the military could produce a Marshall, an Eisenhower, a Nimitz to meet the stresses of World War II.

At the beginning of his second presidential term, Franklin Roosevelt began to mobilize the nation for the war he saw coming when he delivered his "Quarantine the agressor" speech. He recognized that he had not only to build a military force but also to create an industrial base that could supply this force with ships, planes, and tanks. The buildup was a gradual one, made possible by the late entry of the United States into the war. Mobilization had its crises too, as when the prewar draft was renewed by a margin of one vote in the House of Representatives, two months before Pearl Harbor.

Once the war was over, the armed forces rapidly shrank away but not to the prewar levels. By 1948, they were down to 1.4 million men. The Korean War brought about another large increase and a postwar cutback but again to almost twice the prewar levels.

The Eisenhower years were a time in which overall budget constraints imposed by a desire to limit total government spending came into conflict with a relatively aggressive foreign policy—and with the general alarm created by the Soviet Sputnik launch in 1957. Those were the years of the slogans, "Massive retaliation," and "More bang for the buck." Treasury Secretary Humphrey warned of the dangers that excessive government spending could produce by way of a "hair-curling depression," and recognized that military spending was the biggest and most flexible piece of government spending.

With the advent of the Kennedy administration came a dramatic shift in the formula for making up the defense budget. President Kennedy instructed his secretary of defense to meet the security needs of the United States "without regard to arbitrary or predetermined budget ceilings." It was an instruction that no other cabinet officer received about how to wield his responsibilities.*

The defense establishment is not only the largest organization in the United States; it also embraces one of the largest industrial concentrations in the country—employing 2 percent of the United States work force, down from 4 percent in 1968, but still among the largest American industries—and the only one in which the customers all work for the same boss.

Furthermore, it is one of the very few industries whose products offer no direct or immediate consumer satisfaction other than the pleasure members of the armed forces may take in flying airplanes or firing weapons.

Of the total expenditures by the military establishment, more than two-thirds goes for the purchase of plant, equipment, and supplies, as distinguished from compensation for military and civilian personnel.

Half the goods and the great bulk of the services that the military establishment buys are unique to it; the nature of these purchases makes the establishment what economists call a monopsonist. All the big ticket items—missiles, war planes, naval vessels, tanks—have no civilian market (or at least no legitimate one) and most of the smaller items are similarly unique. Only a small portion of military purchases—primarily food, supplies, and construction materials—is produced for a market that is not dominated by the military purchaser. A tabulation for fiscal 1961–1982 appears in table 4.3.

*See chapter 6 below for elaboration.

TABLE 4.3   Department of Defense Expenditures by Category,
Fiscal 1961–1982 *(Billions of Dollars)*

Fiscal Year	Military Oriented[1]	Common to Military and Civilian Use[2]	Subtotal	Pay and Allowances	Total
1961	19.2	10.8	30.0	16.6	46.6
1962	20.9	10.3	31.2	17.8	49.0
1963	23.0	9.1	32.1	18.0	50.1
1964	22.4	10.0	32.4	19.1	51.5
1965	18.0	14.7	32.7	14.8	47.5
1966	20.6	12.2	32.8	22.1	54.9
1967	26.2	16.3	42.5	25.7	68.2
1968	31.0	25.8	56.8	22.0	78.8
1969	31.5	24.1	55.6	23.8	79.4
1970	28.8	24.0	52.8	25.8	78.6
1971	26.2	23.6	49.8	26.0	75.8
1972	25.0	24.7	49.7	26.9	76.6
1973	23.9	23.0	46.9	27.6	74.5
1974	23.8	25.2	49.0	28.8	77.8
1975	24.9	29.5	54.4	31.2	85.6
1976	24.9	32.1	57.0	32.4	89.4
1977	28.0	35.6	63.6	33.9	97.5
1978	30.5	38.4	68.9	36.3	105.2
1979	36.6	42.4	79.0	38.7	117.7
1980	42.1	51.1	93.2	42.7	135.9
1981	50.5	59.2	109.7	50.1	159.8
1982[3]	59.6	74.6	134.2	53.3	187.5

[1] Consists of procurement and RDT&E expenditures for weapon systems and related hard goods.

[2] Consists of all other purchases, including construction, operations, and maintenance.

[3] 1982 figures are estimates.

The consequences of these distinctions for the producer in the military marketplace are several:

A single customer tends to rely on a limited number of suppliers for any specific item or family of items.* This is particularly so when the customer is buying relatively small numbers of very expensive and complicated items: ships, planes, missiles, tanks.

The dividing line between supplier and purchaser tends to be blurred. There are almost no middlemen, nor is there any need for them. In fact, purchaser and supplier employees interpenetrate each other's establishments. When the air force was created as a separate service in 1947, it underwent an enormous and rapid expansion. Top management chose to rely on contractors rather than on civilian employees or military personnel to perform basic development and procurement functions. This decision was not the result of any conspiracy between the elements of the military-industrial complex but was an effort to avoid the nearly impossible task of recruiting officers to perform these greatly expanded functions and an attempt to sidestep the difficulty of persuading Congress to create civilian positions in addition to appropriating funds.

The Defense Department has a considerable inventory of plant facilities that it makes available to industrial contractors to produce defense equipment in exchange for appropriate price adjustments and consequent further involvement of government in the production process.

Complex weapon systems seem to require frequent and extensive change orders after the initial contract is awarded. In one recent three-month period, for example, the navy issued 2,900 design changes for the Trident submarine. To what extent many of these change orders could be eliminated by better advance planning and more efficient management is a question that has been and continues to be hotly debated but, until the enormous problems involved are overcome, the change order process will continue to modify the arms-length relationship between supplier and consumer.

*There are about 25,000 prime contractors and another 50,000 subcontractors in the country engaged in defense-related production at any particular time. Only 33 contractors, however, account for over half of prime contract awards.

In part because of the extensive change order process, competition among suppliers cannot effectively be based on price. In the absence of effective price competition, relationships between producer and consumer are likely to be closer and more subjective and barriers to entry by new firms higher and more difficult to penetrate.

Many reasons have been cited, especially by small enterprises, for not entering the defense market. These include low profits, small volume, one-year orders, cyclical demand, special military requirements, market uncertainty, and excessive regulations and paperwork. Such entry barriers increase the likelihood that small monopolies will develop among highly specialized parts suppliers. This situation can produce expensive delays and uneconomical purchases. For example, Lockheed was forced to buy a 10-year supply of electronic circuits for the P-3 antisubmarine aircraft when it learned that its supplier might quit producing them.*

Lastly, defense contractors are subject to a wide range of regulations, most of which are embodied in a several-inch thick document referred to as "The Defense Acquisition Regulations," that cover every conceivable aspect of management from discrimination in employment to treatment of capital costs. These regulations necessarily involve the military purchaser more heavily in almost every aspect of management decision making than is the case with ordinary commercial purchasers.

None of the above is to deny the existence of vigorous competition within the defense industry. But for reasons already stated and elaborated upon below, the competition takes somewhat different forms than it does in the commercial market. Most of the vigorous competition occurs in pursuit of the initial research and development contract. After this initial competition, the winner may become the sole developer and producer of that particular weapon system for the next 20 years. Not infrequently, the successful competitor first "buys in" by intentionally bidding under expected costs in the belief that subsequent government modifications will provide sufficient slack to absorb unrecovered early costs. Traditional gov-

---

*The problem of the capacity of the United States industrial base to absorb increases in defense spending is discussed below.

ernment thinking has been that, once this initial research and development investment has been made, it is too costly to seek other contractors for subsequent stages of development. As a consequence, some 65 percent of DOD's contracts are awarded on a noncompetitive basis; a practice much criticized by the General Accounting Office.

One of the consequences of these competitive differences is that enterprises operating in the defense sector find it difficult to enter the commercial market and vice versa. A firm that has been engaged in defense production and sees a shrinking market for its defense specialty or greater opportunities to supply civilian customers will not infrequently choose to move into the civilian market by acquiring a subsidiary or by establishing a new division with independent top management. These steps are taken because existing management cannot adjust to emphasizing price competition and volume sales to a wide variety of consumers.

In 1980, twelve companies each did more than $1 billion worth of business with the Defense Department. For five companies, defense contracts constituted over 50 percent of their total revenues: General Dynamics ($3.5 billion from DOD, representing 74% of company revenues); McDonnell Douglas ($3.25 billion, 54%); Hughes Aircraft ($1.8 billion, 59%); Grumman ($1.3 billion, 76%); and Northrop ($1.2 billion, 74%). Additionally, Fairchild Industries, which ranked 24th among defense contractors, did work for the Pentagon amounting to $560 million or 62 percent of the company's revenues.

Defense industry is as customer conscious as other segments of American industry. But this consciousness takes the form of cultivating a single customer—which in turn has bosses in the executive and legislative branches—in ways that are significantly different from marketing in the civilian market.

Defense contractors do not, however, maintain intimate supplier-customer relationships with the Department of Defense as a unit. Rather, they tend to establish relationships with one of the military services, and even with one element within a military service, almost to the exclusion of the rest of the military establishment.* The result is that contractors become deeply involved in interservice and intraservice rivalries, for example, between the submariners and the fleet air arm within the navy. They participate, often actively, in political contests over which weapon system should be chosen and how new increments in the military budget should be shared. Industry lobbyists (both in the Pentagon and on Capitol Hill), institutional advertising, and even substantial (if sometimes reluctant) investments in research and development support their selling efforts. Where the United States is involved in coproduction of weapons with its NATO allies, these selling efforts extend to intensive campaigns to persuade the military to make offshore purchases of an ally's product. Counter-offensives are then launched by entirely United States based contractors to keep as much of the business as possible in the United States. Such actions often frustrate our NATO partners. Similarly, contractors may work with their United States military customers to encourage foreign military sales of a particular item in order to reduce the per copy cost or to keep a production line going. A recent study by Andrew Pierre of the Council on Foreign Relations suggests, however, that economic motives may be less significant than diplomatic ones in these transactions—a matter further discussed in chapter 8 below.

Foreign military sales were first explored on a significant scale as a supplementary market for United States arms in the late sixties when they were seen as a convenient offset to American military expenditures overseas. During the 1970s, military contractors competed more extensively in foreign markets. After a valiant and unsuccessful attempt by the Carter administration to hold down the rate of growth for foreign military sales, the Reagan administration again turned on the spigot full force, revoking the so-called "leprosy letter" by which the previous administration had instructed United States missions abroad not to provide normal commercial assistance to American arms dealers.

This confluence of factors has produced a weapons acquisition

*The army alone, for example, has separate, semi-autonomous R & D commands for armaments, aviation, electronics, missiles, mobility equipment, and tank-automotive matters. Each of these commands does its own contracting. The picture is similar in the other services.

process that, whatever technological wonders it may have wrought, is considered by many knowledgeable observers to be quite inefficient and much less productive than it could be. Perhaps the most extreme manifestation of the process's shortcomings is the rise in unit costs of the weapons produced. Over the past 20–30 years, systems have been getting more costly (in constant dollars) than their predecessors by about 5.4 percent annually. The end result has been a marked decline in the number of weapons per year coming off the assembly line.

For example, after the Korean War, the United States was manufacturing 5,200 tanks per year. Now, only 720 a year are produced. After discounting for inflation, the cost per tank is six and one-half times what it was when production was more than seven times higher. Production of the army's new main battle tank, the M-1, has taken 18 years and has involved three separate designs; this system was considered urgently needed in 1963. In the end, the entire project is expected to cost more than $21 billion, or $3 million per tank, compared with the original estimate of $420,000 each.

The number of fighter planes produced each year since the Korean War has dropped from 5,200 to 440. A modern F-15—priced at $28 million apiece—costs 38 times (in current dollars) the pricetag of an old F-86. The navy's F-18 multirole combat aircraft originally was to have cost about $9 million. Some estimates predict that the F-18 eventually may cost as much as $35 million each.

Since 1957, the number of surface warships has dropped from 355 to 140. A destroyer produced in 1957 would cost about $114 million today. Modern-day destroyers and cruisers cost $1.1-1.2 billion apiece.

Other examples of such "gold plating" abound. Probably the most notorious example is the Trident submarine. Originally expected to cost $900 million apiece, the Trident now is priced at $1.2 billion per copy and is more than two years behind its production schedule. The army's new infantry fighting vehicle has doubled in price from $900,000 to $1.8 million each. SOTAS, a helicopter-borne radar system, has more than tripled, from $8.7 million to $28 million per unit.

These examples bear out an observation by Thomas R. Stuelpnagel, former president of the Hughes Helicopter Company: "One of the first myths about hardware is that its cost can be controlled. The fact is that once the hardware is designed, and especially in production, the hardware is in control."

If unit costs continue to rise as they have done in the recent past, procurement spending must increase accordingly just to stay even. If procurement spending does not keep pace with unit cost growth, then either the age of the weapons inventory must increase or the inventory must decline. James Wade, the current deputy under secretary of defense for research and engineering, has demonstrated that, if the annual procurement growth were 5.4 percent and unit cost growth 2.7 percent, inventory could be increased about 70 percent over 20 years. On the other hand, if unit cost growth continues at the historic rate of 5.4 percent and procurement growth stays at 2.7 percent, then the inventory must decline by 42 percent over the same 20 years.

The Reagan administration's response to this problem has been to increase procurement spending, from roughly $45 billion (slightly over 26 percent of the defense budget) in 1981 to about $65 billion (32 percent of the defense budget) in 1982. At the same time, Defense Secretary Caspar Weinberger has set a goal of saving $28 billion by 1985 through increased competition, streamlined supply systems, and the elimination of cost overruns. There is some skepticism about his ability to meet this goal.

In addition to the impact of the military establishment on the behavior of individual supplier firms—that is to say its microeconomic effect—we need to examine its macroeconomic impact on the United States economy as a whole. That impact can be characterized as generally inflationary, sometimes countercyclical, or at least independent of the business cycle, and probably taking as much from the technological development of the civilian economy as it contributes, although that last judgment is a difficult one to make. In 1982, the military establishment cost the average American adult about $1,250.*

Economists generally agree that when, in the mid-1960s, Lyn-

*This figure is based on a 1982 budget of $199.7 billion spread among the 158 million Americans indicated by the census to be over age 18.

don Johnson insisted on pursuing his Great Society programs while greatly expanding the scope of the Vietnam War, asserting that we could have guns and butter too, he launched the American economy on the first wave of the inflationary surge that we are still struggling to overcome.

Unlike spending for the production of food, clothing, shelter, or gadgets, defense spending, without compensatory tax increases, is inherently inflationary because it does not add to the sum of goods and services available to consumers, however essential it may be to national security. When the widget factory expands its production, the additional dollars that go into the pockets of its employees and suppliers are offset by the additional widgets that those additional dollars can purchase. Not so with national defense.

There are four additional reasons why defense spending tends to be inflationary:

First, because of the special characteristics of the defense market, the skills and capital required, and the consequent barriers to entry and exit, as discussed above, a rapid increase in the defense budget tends to raise the prices of existing suppliers rather than to attract new suppliers into the market.

Second, because defense production involves a relatively high proportion of skilled labor (except for shipyards, where there may be large numbers of entry-level jobs), and because that labor generally is highly organized and can obtain automatic wage increases that feed the inflationary spiral, the inflation rate for defense industry tends to be higher than for the economy as a whole, even in times of normal growth.

Third, since competition in defense industry is based primarily on quality rather than on price, customer resistance to price increases is less than in the civilian market.

Finally, as difficult as it is to measure the incidence of uneconomic spending decisions, it nonetheless seems incontestable that a relationship exists between such decisions and inflation. One of the inevitable consequences of a pluralistic political process is that decisions inconsistent with economic prudence often emerge. Where waste results, inflation is fed by the costs of opportunities that must be foregone in the near term but paid for in the long run. Two kinds of decisions, in particular, typify this genre:

The first involves the cancellation of a weapons program that has been ongoing for several years and, in the process, has absorbed a great deal of money. The Reagan administration's decision to kill the army's Roland air defense missile program, despite the fact that more than $1 billion had been sunk into it, is a case in point. The particular decision to terminate the program may be wise or foolish. Yet the cumulative effect of programs terminated after heavy costs have been incurred but before any benefits to the force structure have been realized cannot help but be inflationary.

A second kind of decision is to continue purchasing weapons that are no longer needed (or wanted) by the military. This situation is most likely to occur where termination of a contract threatens the closure of a plant in an economically vulnerable community or in an influential congressman's home district. For example, after original plans to stop funding the P-3C patrol plane threatened to halt Lockheed-California's production line, funding for the purchase of two or three more of the aircraft may be requested (and obtained) from Congress in the FY 1983 budget.

A special inflationary impact on the economy results from the military establishment's heavy demands on United States energy supplies. The military has made extraordinary efforts to curtail its consumption of energy but such reductions are limited by the primary needs for transportation and training, as well as for subsidiary uses. The statistics are impressive. The military accounts for roughly 65 percent of federal energy consumption (but only about 1.5 percent of the overall national figure, down from 2 percent in 1975). It is by far the dominant consumer of fuel in the federal sector. Excluding purchases for the Strategic Petroleum Reserve, the Defense Department has accounted for an average of about 95 percent of federal fuel purchases over the last four years. Approximately 85 percent of the military's petroleum consumption is for aircraft, ship, and ground operations; only 14 percent is used for housekeeping. In 1978, fuel outlays were 3 percent of the overall defense budget; by 1981, that figure had doubled.

When plans for deploying new missiles coincide with other governmental and private efforts to develop the same regions of the country for extraction of coal and shale, the environmental impact becomes no less a major problem than energy consumption. The political furor generated by potential environmental impacts undoubtedly played a major part in President Reagan's tentative decision to house MX missiles (when they are fielded) in existing missile silos. This plan differs significantly from the "racetrack" scheme originally proposed by the Carter administration. The original scheme would have involved shuttling 200 MXs among 4,600 shelters over roughly 8,000 miles of roadway and almost 6,000 square miles of territory in Nevada and Utah. Construction of such an arrangement would have required approximately 600,000 tons of cement, 32 to 48 million tons of sand, 210 million gallons of liquid asphalt, 125 million gallons of petroleum fuel, and 17.9 billion gallons of water.

None of these problems is limited to the defense sector and all of them contribute to the general inflationary tendencies of the United States economy. They are, however, strongly characteristic of defense spending.

The countercyclical character of defense spending, or at least its independence from the general business cycle, is self-evident. Defense production does not rise in response to a general increase in the level of consumer demand, nor does it fall because the general business climate is unfavorable. Indeed the Reagan administration's heavy government spending in the defense area seems to run directly counter to its domestic economic policy and may prove to be the Achilles' heel in its overall economic program. Budget deficits totaling $423 billion have already been forecast for fiscal 1982–1984. The Reagan program of large increases in the defense budget accompanied by significant tax cuts could, in the words of MIT economist Lester Thurow, "wreck the economy" by bleeding key human resources, scarce materials, factories, and financing from other sectors. The 1982 defense budget could be said to have been purchased at the expense of cuts in other programs, including school lunches, job training for the poor, food stamps, low-cost housing, health care, financial aid for college students, and local government grants. While it might be argued that these cuts would have been made in any event, it seems clear that the current economy would have great difficulty accommodating anything like the 1980 level of social programs, together with the Reagan defense budget, even without the Reagan tax cuts.

An even deeper concern is the impact of defense spending on the availability of investment capital to modernize civilian production facilities. In a period when economic revival is agreed to be a captive of too little spending on industrial (and postindustrial) development, rather than too little demand for consumer products, diverting investment dollars and energies into military production is not likely to help pull the economy out of its depressed state. Nor will military spending contribute to lower the interest rates that block new private investment in the civilian economy. By competing for scarce investment dollars, and by increasing the federal deficit, defense spending tends to keep interest rates high.

The general state of the economy may put a brake on military spending, as it did explicitly in the Eisenhower administration. The Eisenhower restraint was designed to avoid an accidental coincidence of higher military spending with a possibly overheating economy. There is no recent recorded instance of defense spending being used deliberately to bolster a sagging economy in the United States, although it generally is believed that it was defense spending that gave the decisive push to get the country out of the Great Depression of the thirties.

In fact, there is a growing body of opinion today that increases in defense spending can do more to depress than to bolster the economy. In the words of Nobel Prize winning economist, Wassily Leontief: "If handled improperly, these huge jumps in military spending will mean higher inflation, a worsening balance-of-payments gap, a drain on productive investment, soaring interest rates, increasing taxes, a debased currency and, in the longer term, more unemployment . . ."

The results of one recent study by Michigan researcher Marion Anderson suggest that, contrary to prevailing belief, military spending actually generates unemployment. According to Anderson, every billion-dollar increase in the military budget costs America some 10,000 jobs. If people all over the country are paying high taxes, a substantial percentage of which goes to the Pentagon, they do not have control over that money. As a result, they build fewer houses, buy fewer cars, take fewer vacations, and vote lower taxes for state and local governments. Anderson's figures show that $1 billion spent in retail trade creates 65,000 jobs. The same amount spent on education produces 62,000 jobs; on hospitals, 48,000 jobs; but, on guided missiles or ordnance, only 14,000 jobs.

Of equal, if not greater, concern is the question whether the United States industrial base is capable of absorbing very large military spending increases. The major source of this concern is the second-tier subcontractors who supply parts and raw materials to the big assembly lines and shipyards. Although about 50 percent of defense spending goes to subcontractors at all levels, small firms have been opting out of defense work in increasing numbers. According to one estimate, the number of companies involved in aerospace production alone has declined by more than 40 percent since 1967, falling to about 3,500 from about 6,000. The average age of some of the critical items of machinery within the machine tool industry is 27 years. The backlog of orders in that industry is $5.2 billion, and lead times for some castings and forgings may be as long as 80 weeks. Forging companies are working at about 60 percent of capacity. The net of this state of affairs is that the previous pattern and process of military spending may have reduced the defense-oriented portion of the machine tool industry to the point where it cannot be fully responsive to present military demands.

Another general effect of military spending on the economy is that military research and development have made significant contributions in such areas as the development of radar (a by-product of World War II), the jet aircraft engine, micro-miniaturization of computers and other electronic tools, and the practical uses of laser energy, to name only a few of the most dramatic examples. Absent the concentration of massive resources on a limited research and development objective, these innovations, or at least their practical applications, might have been much longer in gestation had they been dependent on private or public nondefense funding.

But, as staff members of the Council on Economic Priorities noted recently, two problems increasingly plague defense-related innovations. First, America's record in translating military advances into competitive civilian products has been less successful in the past decade. For example, although solid-state circuits were an innovation financed by the United States military, it was the Japanese who used them successfully to decrease costs of television sets, stereos, and videotape records. Second, the fact that the military tends to emphasize technical performance over cost makes most defense technologies inappropriate for commercial applications. As an example, military demands are pushing American development of computer chips in the direction of higher speeds at higher costs, while Japanese manufacturers are developing cheaper, more reliable chips with greater storage capacity. Ultimately, these problems could (1) make the United States less competitive in the international commercial arena; (2) diminish America's role as a leading force in technological innovation; or (3) make the defense sector an even less desirable arena for competition by American industry than it is at present.

In the final analysis, the negative effects of military research and development on the civilian research and development process probably still outweigh the positive. One need only ask what would be the predictable reaction of military planners if they were told they would have to rely exclusively on a bunch of "damn civilians" to supply their needs.

The extent to which the output of scientists and engineers in the United States has been appropriated by the Department of Defense is quite staggering. Conservative estimates indicate that defense and space programs employ 20 percent of all American scientists and engineers engaged in research and development work. Other estimates go as high as 50 percent.

Since engineers in particular have been in short supply recently, one can argue that if fewer of them had gone into the defense area more of them would have been available for civilian research and development projects. Further, the predilection of the Defense Department for dealing with relatively large civilian business organizations may deflect scientists and engineers away from the smaller enterprises where the seminal ideas in industrial research often originate. It is worth noting that the shortage problem does not end with scientists and engineers but extends to the likes of machinists and tool-and-die makers. Estimates are that the defense industry will need 10,000 machinists and tool-and-die makers by 1985 just to make up for attrition and to maintain current levels of production.

On the other side of the ledger, university research is now, more than ever, heavily dependent on Defense Department support, particularly for the kinds of basic research (in high-energy physics, for example) that require large capital investment. Defense Department support of university-based research amounted to $639 million in 1982, up by 42 percent since 1980. With this increase, the Pentagon controls 53 percent of the total federal allotment for research and is the source of 13 percent of all university research funding. Defense Department support is particularly important not only for the universities but for the economy as a whole because the universities are the principal sites of basic research. Basic research, which is the seedbed of industrial development, has been suffering from increasing neglect over the past several years. Less than four-tenths of one percent of the entire Defense Department budget goes for research, and 76 percent of that is devoted to applied research.* But such figures, while small relative to overall defense spending, nonetheless represent a major research funding source for universities. Consequently, with traditional funding sources drying up, the universities are turning more and more to the Pentagon for support. In 1980, there were twelve universities that each did over $10 million of business with DOD, and two of those—Johns Hopkins and MIT—received over $150 million each. A decade ago, with the antipathy generated by the Vietnam War at its height,

---

*Of the $15–20 billion annually appropriated for "Research, Development, Test and Evaluation," more than 95 percent goes for "Development, Test and Evaluation."

university funding was at its lowest ebb; it was during this time that the Mansfield Amendment (sponsored by former Democratic Senate Majority Leader Mike Mansfield) limited defense contracts to work that had a "direct and apparent relationship to a military function." This mandate was considerably softened by amendment in 1970, the year after it was first enacted. Thus, whether by fiat or natural inclination, the Pentagon-university relationship fell into serious disrepair. It only now seems on the verge of recovery.

No account of the impact of the military establishment on the economy would be complete or adequate without some discussion of the reverse impact of the defense industry on government, beyond what is described in chapter 3 above. The intimate relationship between government and the defense industry cannot help having its effects on both factors in the equation. That this close relationship exists is not to suggest that there is a conscious conspiracy or series of conspiracies against the public interest but rather that as the suppliers tend to accommodate to the desires and habits of mind of their single customer, so the customer tends to adapt to and even adopt some of the habits of mind of its suppliers. This tendency is encouraged by the fact that many retired military officers find their way into jobs in defense industry where their expertise as well as their acquaintanceship will be valued. Serving officers often find themselves dealing with old acquaintances or former patrons from the service and they cannot ignore the prospect of moving across the bargaining table when they retire—particularly since the military personnel system favors earlier retirement than is common in civilian life.

The enthusiasm of industry engineers for a new weapon system they have designed or helped to design is contagious. The same is true of their concern over the possibility that a design team will be broken up because the contract is not renewed or that a production line will be closed down because the service is curtailing its deployment plans. It is only a step from coordinating design efforts for a proposed weapon system to coordinating lobbying efforts on the Hill to find the funds to carry on the design program and then to take it from development to production and deployment. Long-established relationships between military men (and civil servants) on the one hand and defense contractors on the other may survive generations of political appointees and reach over the heads of those appointees to congressional committee members and staff. If the net result is to make defense spending and the defense portion of the economy less responsive to political control, this can scarcely be surprising. It would be strange indeed if a symbiotic relationship as close as the one described here did not provide benefits for both parties quite apart from, and not always consistent with, benefits to the Republic.

V

The Impacts on American Society

THE SOCIAL IMPACT of the military establishment is felt directly by the 130 million Americans who have served in uniform, in peace and in war, since 1940 and by the 350,000 who enter the service every year, the great majority for a two- or three-year tour. For our present purposes, we will not consider the numbers and the consequences if the United States gets into another shooting war. But the very existence of the military establishment also has a significant social impact even on those who have not served and who do not expect to be called upon to serve. How the second—and very much larger—group is impacted depends in part upon how the first group is selected: who serves and who does not serve.

Those serving in the United States military today are significantly different from their civilian contemporaries, although worsened economic conditions seem to be narrowing the gap. Enlistees are generally less well educated; they score lower on tests designed to measure basic mental abilities; and their families appear to be less well off. These deficiencies contrast sharply with the quality of recruits who entered the service as recently as the 1960s. To complete the profile, present-day enlistees are more likely to be black, and more than 90 percent are male.

This chapter has benefited particularly from the input of James Skelly, whose assistance is gratefully acknowledged by the authors.

On the other hand, high civilian unemployment, a protracted recession, better pay, and (some would say) the revived prestige of the military contributed to an interesting reversal of trends in 1981 that proponents of the all-volunteer force hope will be a portent of things to come.

Individuals enlisting in the armed services in 1980 had an educational level significantly lower than that of either their civilian contemporaries or of those who entered military service in the final years of the peacetime draft before Vietnam (i.e., before 1965). The number of high school graduates was markedly lower in 1980, and very few men and women with some college education enlisted. By the end of 1981, all of this had changed appreciably, so much so that Senator Sam Nunn, a Georgia Democrat and a leading member of the Senate Armed Services Committee, was prompted to observe: "Supply-side economics is working . . . It is supplying the military with volunteers."

Prevailing trends tend to be most evident in the army, which historically has had the greatest need for conscripted service. The percentage of army enlistees with some college experience dropped markedly from 14 percent in 1964 to 3 percent in 1979. By 1981 the figure stood at 9 percent (which compared with the 16 percent of the entire male population that had some college experience). Similarly, 67 percent of those entering the Army in 1964 were high school graduates, a figure that dropped to 54 percent by 1980 but rose to 80 percent in 1981.* Since the high school diploma generally is accepted as the best indicator of an individual's ability to adapt well to military service, the educational level of enlistees will continue to command special attention. Generally speaking, a high school graduate has an 80 percent chance of successfully completing a three-year term of service, while a non graduate has only about a 60 percent chance. In recent years, fully one-third of those who have entered the service have failed to complete their initial obligation because of job inaptitude, indiscipline, personality disorders, or similar factors.

*The number of diploma-holding enlistees in both the navy and the Marine Corps increased approximately 19 percent between 1964 and 1981, while the figure increased about 4 percent during the same period for the air force.

A related concern has been the proportion of recruits in the upper and lower mental categories. The proportion of military entrants in the top mental categories (I and II) fell from 38 percent in 1964 to 27 percent in 1980 but rose to 33 percent in 1981. Similarly, the proportion placed in category IV, the lowest the military can by congressional direction accept, rose from 14 percent in 1964 to 31 percent in 1980, then dropped to 18 percent in 1981. Within the army alone, 1980 enlisted accessions included 50 percent category IV personnel, a figure that dropped to 31 percent in 1981.*

The results of intelligence and educational deficiencies show up most clearly in the ability (or inability) of the military to operate and man today's highly sophisticated technological weapon systems. One army study, for example, demonstrated that more than a fifth of the tank gunners serving in Germany (and more than a quarter of those serving in the United States) in the late 1970s did not understand the procedure for aiming their battlesights. Soldiers in mental categories III and IV often have difficulty understanding instruction manuals, and, as the study points out, are therefore unable to operate the sophisticated components of the army's air defense systems or even such basic individual weapons as the Redeye missile. In the case of the Redeye, soldiers often were not able to recall the firing sequence, nor were they able to differentiate between enemy and United States aircraft; this in spite of the fact that a "high percentage" of Redeye crews were drawn from category III rather than from category IV.

The criticality of such findings is obvious. As weapons become increasingly sophisticated, the aptitudes of weapon operators become the object of unremitting concern to military planners. The new Black Hawk helicopter, for example, has 257 knobs and switches, 135 circuit breakers, 62 displays, and almost 12 square feet of instruments and controls. Considering that the army plans to introduce close to 200 new systems of all types within the next 10 years, the magnitude of the problem is only beginning to be felt. Technical requirements in the air force are even more demanding,

---

*Mental category IV consists of those individuals whose scores on standardized service tests place them in the 10th to 30th percentile of the potential recruit population, as measured against soldiers tested during World War II.

there being almost as many officers in scientific, engineering, and technical billets—17,140—as there are pilots—18,602. As of January 1982, there were 223,884 air force enlisted people in technical jobs, compared with 11,000 who flew in planes.

There is a special concern about the percentage of blacks and other minorities in the force. Minority personnel constitute an increasingly large percentage of males entering the military for the first time and of overall strength. Forty-one percent of the army's overall strength during 1981 were minorities. While blacks as a group tend to be overrepresented in today's volunteer force (they constitute 15 percent of the United States youth population), they were underrepresented before the end of the draft. In 1964, for example, black enlisted men made up about 12 percent of the army and only 6 percent of the navy. By way of further contrast, they represented a mere 7½ percent of enlisted strength shortly after World War II.

The increasing numbers of minority personnel in the military, coupled with what appears to some to be underrepresentation of urban and suburban middle class whites, make the situation volatile. Senator Nunn has expressed this concern clearly with an eye to the potential political consequences:

> We will have a real problem if we go to war, with the combat fatalities that inevitably are going to occur disproportionately in some groups. We are going to have a massive social problem because while the armed forces are viewed as a good job opportunity, in wartime that becomes an opportunity for death on the battlefield. You are going to have an understandable cry from minorities as well as those who don't believe that people lower on the economic ladder should bear a disproportionate amount of wartime deaths and casualties.

The social composition of the military may also provide some indication of how well we are achieving general social goals. Sociologist Morris Janowitz has argued that, "A free and democratic society has to have a representative military" because it "is the basis of civilian control and the legitimacy of the military" and " . . . is essential to military effectiveness." The effectiveness argument has been elaborated by Janowitz and Northwestern University sociologist Charles Moskos. They comment that a military force that is disproportionately high in minority content may heighten racial tensions and cause severe management problems for the military. Further, it may lack credibility with other nations and ultimately with American whites. They worry that a "tipping point" may be reached where whites will be further discouraged from joining since either the armed forces as a whole, or particular branches of the service, could then be perceived as a minority military force thereby defeating the objective of large-scale integration.

Unfortunately, because of the difficulty of obtaining such figures in the first place, hard data on the relative economic status of the families of today's volunteers are not readily available. Nonetheless, it can be inferred from periodic surveys conducted by the Defense Department that, in the aggregate, today's volunteers are more disadvantaged economically than their counterparts who elect not to join the military. For example, youth who indicate a positive propensity to enlist typically have completed less formal education, have lower grades in high school, have fathers with less formal education, and are less likely to be white and employed than those who indicate a negative propensity to enlist. Furthermore, when offered an incentive package consisting of a $3,000 cash bonus, tuition assistance of $2,000 a year, and a $10,000 low interest loan, two-thirds to three-quarters more enlistment-age youth express a positive enlistment propensity than otherwise would have been the case in the absence of such inducements.

These characteristics are not atypical of armies throughout history. When the Duke of Wellington was asked on the eve of the Battle of Waterloo whether his troops frightened the enemy, he is said to have replied: "I hope they do. By God, Sir, they frighten me!" Even the Vietnam draft impacted more heavily on the children of the less privileged because of a system of educational and other deferments that was abandoned only toward the end of the Vietnam war.

In industrialized, urbanized, upwardly mobile societies, it is not surprising that military service seldom appears as the preferred entry route to civilian work and civilian careers. Yet a tour of military service can inculcate basic skills that the young person has not acquired in school—skills in punctuality, responsiveness to instructions, neatness, and civility—as well as a fundamental self-confidence that comes with success in a working situation. Studies indicate that young people of similar capability and background who experience a term of military service do better, on average, in jobs and earnings than their counterparts without such service.

Whatever the pluses and minuses of military service for today's recruit, his/her situation differs significantly from that of earlier generations since the 1950s, and indeed (with brief interruptions) since 1940: his/her service is voluntary. Enlisting may, in many cases, seem the only alternative to continuing unemployment and poverty but it is today a voluntary act. By the same token, military service is not generally regarded as an obligation of citizenship. It is an experience as alien to the majority of civilian youth as being poor, black, or a high-school dropout. The situation is not unlike that described almost a century-and-a-half earlier by Alexis de Tocqueville: "The best part of the nation shuns the military profession because that profession is not honored, and the profession is not honored because the best part of the nation has ceased to follow it."

Peacetime conscription was first instituted in 1948, and abandoned after the United States withdrawal from Vietnam in 1973. In 1980, draft registration was reinstated and over 91 percent of the country's 18-year-old males complied with the order to register. By the second year of registration, however, only 77 percent of those eligible signed up. President Reagan, who as a presidential candidate had campaigned against registration, calling it unnecessary, publicly contemplated calling it off. He temporarily embargoed prosecutions for failure to register but later changed his mind. Today, with a backlog of more than 500,000 nonregistrants, the first exemplary prosecutions have begun.

Because the peacetime military can use only a fraction of those who turn 18, the system has a kind of built-in unfairness that tends to be exacerbated by whatever system of preferences or priorities is developed. Politically, a return to the pre-Vietnam peacetime draft seems unlikely. President Reagan has pledged not to restore it, arguing that improved pay, benefits, and working conditions can lead to a workable volunteer force.

There is, however, a revival of interest in alternatives to the all-volunteer idea, primarily because of the reduced size of the 18-year-old cohort in the 1980s. In 1978, some 2.14 million American males reached age 18. In 1980, the figure declined to 2.13 million and by 1995 it will fall to 1.66 million. Not until the year 2000 will the 18-year-old male cohort rise to 1.99 million; if one assumes that roughly one-third of those will be ineligible for service for medical, intelligence, or moral reasons, or because they are medical or dental students, then the pool of "availables" in that year will be only 1.33 million. Further exacerbating the problem is the congressionally-mandated ceiling on the category IV recruits that each service can accept. In FY 1982, each service was allowed a maximum of 25 percent category IVs among its new personnel, while from FY 1983 on, the figure is set at 20 percent.

For the army in particular, this situation was expected, before the 1981 results were in, to create major recruiting difficulties for the future, especially in light of Reagan administration plans to increase the size of the force by almost 250,000. Ironically, the army only recently has begun to achieve its annual recruiting objectives. It now has done so two years running. On the other hand, it has experienced some significant shortages of experienced personnel in critical skill areas. Similar difficulties have confronted the navy and the air force. The navy, whose 450-ship fleet is scheduled to expand to 600 by 1990, is short of 22,000 petty officers; the air force must expand by almost 50,000 personnel in order to meet new requirements—a new manned bomber, support of the Rapid Deployment Force, the deployment of ground-launched cruise missiles, and space operations. Over the past four years, 12,000 pilots and 5,000 navigators have left the air force for commercial aviation jobs. Indeed, the availability of volunteers may impose a more effective ceiling on expansion than the availability of dollars, particularly since new weapon systems seem to have a priority claim on dollars. Even though 1981 saw air force pilot retention rise by 30 percent and noncommissioned officer reenlistments increase by 12,000 across the four services, such short-term trends rest on the most tenuous of foundations. Only time (and the state of the nation's economy) can tell how lasting these trends will be.

In the current debate, a number of proposals has been advanced to increase the quantity and quality of recruits. These proposals do not, of course, go to maintaining the officer corps, which depends heavily on ROTC for the bulk of its young officers and equally heavily on the service academies for most of its long-career officers. Army ROTC enrollment doubled between 1974 and 1981. During the same period, air force ROTC enrollment increased 25 percent, while the navy program accepted only 6,000 candidates in 1981 from among 30,000 applicants.

In view of impending personnel crises in the services, debate has focused on four possible solutions:

Universal military service, a recurrent proposal, has been suggested primarily as a demonstration of national political will and a way of avoiding the inequities of selective service, despite its very considerable cost and the lack of any clear military justification.

A *renewed selective draft* has been proposed, with minimal deferments or exemptions. The student deferments which distorted the Vietnam draft would be abandoned, and even the physically unfit might be required, if selected, to do limited military or substitute civilian service. It is by no means clear that a selective draft, even if politically acceptable, could be operated in a way that is economically and socially neutral, i.e., that calls up the same proportion of upper class whites and poor blacks.*

If a draft were used only to make up the shortfall between force requirements and voluntary enlistments, it is not clear that it would significantly change the social and economic composition of the military. If young men were not permitted to volunteer, it might be

---

*It has been noted that a return to the draft, in any form, would again raise the old question of "Who serves, when not all serve?" One of the legitimizing factors of the 1950s-era draft was the combination of a small national youth cohort and the large size of the active force. As a result, over three-quarters of eligible men served in the military. Under present manpower requirements, only about one male in five would serve, and even with the manpower increases planned by the Reagan administration the ratio would be unlikely to exceed 25 percent.

argued that an important avenue of upward social mobility was being denied to those most in need of it. Also, while the constitutionality of excluding women from the draft has been established by inference in the 1981 Supreme Court decision validating a male-only registration program, a system that excluded women from serving even as volunteers would create a considerable outcry. A system that permitted women but not men to volunteer might also be subject to serious criticism.

Lastly, some have argued that a military made up primarily of draftees would be less likely to be employed in military adventures and would be more subject to intensive public scrutiny. The power of the draftee's mother is not to be overlooked by any politician; somehow the mothers of volunteers are not in quite the same strategic position as the mothers of draftees.

A third option would be *a national service program* for all young people, in which registrants might be given the option of military or civilian service. Such a program could be either compulsory or voluntary, but if the latter, on a very large scale, with a recruiting program designed to bring in perhaps one million young people a year.*

Interestingly, the national service concept is not a new one, although its potential utility commands more attention today than in the recent past. In 1906, William James published his famous essay, *The Moral Equivalent of War*, in which he proposed conscription of the entire youth population for nonmilitary service. Seeking a new civic discipline for each generation to equal the patriotism, toughness, and initiative normally reserved for war, James argued that no democracy can survive unless a strong civic virtue is imprinted upon each new generation through nonmilitary service for the common good.

A massive but voluntary national service program could restore the idea of voluntary service, while providing special benefits to those who choose the more arduous military service options.

---

*It should be noted that even cursory estimates of various national service options place the costs of such programs at anywhere from several billion dollars to as much as $25 billion a year.

Another feature of the voluntary program might be a standby draft from among those who do not volunteer for either civilian or military service. The standby draft would be invoked only if military volunteers do not fill the ranks.

One aspect of national service proposals is the question of opening military service to a significantly larger number of women. The number of women on active duty in the armed services increased more than fourfold with the advent of the all-volunteer force, from less than 2 percent at the end of fiscal 1972 to roughly 9 percent in 1981, with a planned increase to more than 12 percent by 1986. The Reagan administration, however, instituted a freeze of this percentage pending detailed review of gender-related policies and programs by each of the services.

A number of legal restrictions still exist as to where and how women can serve; some policies differ among the services but all of them are essentially designed to exclude females from combat roles. The traditionalist arguments are that women interfere with male bonding in combat units and that the natural instinct of male soldiers to protect women could get in the way of performance of their combat duties. Conversely, it is argued that full integration of women would overcome outdated male stereotypes inconsistent with an egalitarian society. The parallel with earlier military attitudes toward integration of blacks has been noted by at least one observer.

Despite all these restrictions, the military has made enormous strides in the integration of women into the traditionally male military society. Over the last ten years the separate Women's Army Corps and women's units in the navy and marines have been abolished. Women have been admitted to the service academies; they have been given flight training; and they have been assigned to missile crews, to noncombatant ships at sea, and to an army airborne division. Today, 92 percent of army skills are available to women. In 1972, 90 percent of the enlisted women on active duty in the army were found in traditional (administrative and medical) jobs. Over 60 percent of the army's active duty women now perform nontraditional jobs (maintenance, law enforcement, ammunition, military intelligence, etc.).

A final option would consist of *new incentives to encourage recruiting*, with emphasis on educational benefits that might attract the kind of volunteers who were less motivated by immediate financial rewards. It has even been suggested that existing financial aid programs for college students be made dependent on voluntary service. Such assistance could be a very attractive carrot for participation in various forms of national service. The rewards might be made greater for those completing a two-year stint in the armed forces than for those working in civilian service programs. Clearly, the reinstitution of a more substantial GI bill could serve to motivate significant numbers of able young men and women to volunteer for military service, since studies indicate that this inducement attracts better quality volunteers than do cash bonuses.

This last proposal, advanced most articulately by Moskos, rests on a more fundamental distinction that he draws between military service as an occupation and as a calling. Moskos argues that the real problem with the all-volunteer force has been the manner in which it has redefined military service. He observes that one of the enduring effects of the 1970 Gates Commission's recommendation to make monetary compensation for military service comparable to civilian pay has been to redefine military service as an *occupation* as opposed to what it traditionally has been, a *calling*. He defines a calling as a pursuit that "is legitimated in terms of institutional values and norms, i.e., a purpose transcending individual self-interest in favor of a presumed higher good." Compensation for those involved in a calling is not only monetary but also in benefits such as high self-esteem resulting from special service to the larger society. Moskos notes that military service has had many of the traditional characteristics of callings:

> extended tours abroad; fixed terms of enlistment; liability for twenty-four-hour service; frequent movement of self and family; subjection to military discipline and law; and inability to resign, strike, or negotiate over working conditions.

Thus, in order to serve the needs of something that transcends himself (or herself), the military man (or woman) largely renounces narrow self-interest in return for the protection and care of a predominantly paternalistic institution.

The occupational model defines military service not as a calling but in terms of its economic value relative to other forms of work. The result, according to Moskos, is that "the occupational model implies that priority inheres in self-interest rather than in the task of the employing organization." Thus, he argues, the nature of military service is demeaned, with the result that the services have difficulty attracting from a cross section of society people with sufficient dedication to the institution.

As far back as the American Revolution we see the beginnings of a tie between military service and citizenship. It is a tie that remained more or less intact until the period immediately following World War II. Military service, although conscripted in three wars, was perhaps the most important of the normative aspects of citizenship. Nationalist symbolism was used to legitimate the leadership of revolutionary America and it served as the focal point around which the military forces were organized. Military service has also served as a catalyst for the political ambition of men whose military experience brought them closer to identification with the national community. For example, in the 88th Congress, 66 percent of the members had had some military experience. Furthermore, for the individual soldier, the legitimacy of the armed forces was forged through an ideological appeal to defend individual liberty linked to the achievement of a just society. Thus arose the prototype of the citizen-soldier.

The institutional model began a steady deterioration, however, following World War II and the first use of nuclear weapons. Technological innovations signaled the declining significance of the individual in the prosecution of modern war. This trend was reinforced by the near total negation of the "heroic" image, which had been implicit in feudal military organizations and had survived into the nineteenth century, as the officer corps made the shift from acting as protectorate of the crown to guardian of the democratic nation-state. The heroic image came to be displaced as military organization grew more complex, assuming bureaucratic forms that came to dominate the military social order.

The citizen-soldier ideal also has been eroded by the increasing convergence of civilian and military lifestyles. Officers and NCOs now tend to work more in offices than in command posts and to live away from military installations in housing developments, alongside their civilian counterparts. The base commander attends the local Rotary lunches. Even in the lower enlisted ranks, the bachelor (or spinster) life in the barracks is no longer the rule.

The decline of the heroic image and the convergence of civilian and military values are inextricably related phenomena, the former being subsumed largely by the latter. British journalist Henry Fairlie has gone so far as to describe America as a country lacking the ability to produce, or even to acknowledge, heroes. He observes: "A society that has no heroes will soon grow enfeebled. Its purposes will be less elevated; its aspirations less challenging; its endeavors less strenuous. Its individual members will also be enfeebled." Thus, if the heroic image has lost its hold in the military, it is due in large measure to a more pervasive societal trend.

All these changes coincide with the rise of the strategy of deterrence, which puts the individual soldier at one more remove from the attainment of clear-cut military objectives. A limited war, like the Korean War, where men are asked to risk their lives for limited objectives, really stretches the ties of commitment. As Samuel Huntington points out in his discussion of the Korean War, "For the first time in American history the common soldier fought a major war solely and simply because he was ordered to fight it and not because he shared any identification with the political goals for which the war was fought." A war like the Vietnam War, that divides the country and eventually loses the support of a majority of the American people and its civilian leadership, tests commitment to the breaking point, with resultant fraggings, drug abuse, and atrocities.

In addition to the effects of deterrence strategy, the technological advances in weaponry, and the declining significance of the

individual soldier, Janowitz cites other factors that have contributed
to the diminution of nationalism as an organizing focus for the armed
forces:

> It is also a popular expression of an advanced industrial society with a
> high level of income and education that produces strong antiau-
> thoritarianism and pronounced hedonistic concerns. Negative at-
> titudes toward military service become widespread as the logic of
> nationalism is questioned and as the rationale for the military is ob-
> scured by the reality of nuclear weapons."

Thus, it appears that the problem is not only as Moskos asserts,
that we need "to break the mind-set that sees the all-volunteer force
in terms of economic models." The use of marketplace standards to
man the military forces reinforces long-term trends both in the
society-at-large and in the strategic imperatives of the United States
political and economic position. The attitudes of current recruits are
not encouraging. In a piece by Michael Daly in the editorial pages of
the *Los Angeles Times* in late July 1980, Daly interviews William
Sander Vasquez, who enlisted in the Army in 1966 to escape his
continued conscription into a life of poverty in the ghetto of New
York:

> "Why do people enlist?" he was asked. "Two basic reasons: Food and
> shelter." "What about patriotism?" he was asked. "Not lately," Sander
> said. "There was a guy from the Dominican Republic who said he
> owed America. That was two years ago."

The occupational model of military service was further en-
trenched by attempts during the mid-70s to unionize the armed
forces. The abortive unionization effort pushed by the American
Federation of Government Employees (AFGE) provided evidence
of the strength of the occupational model and, in the process, high-
lighted an underlying conflict in personnel management and in the
definition of military service.

The AFGE had represented many of the civilian workers at
military installations in the United States. Concern with their
financial well-being led the union to attempt to enlist rank-and-file
military personnel. In 1974, the leadership of the AFGE made its

first overture, albeit an indirect one, to service personnel. For the fifth time in as many years, AFGE's civilian members were faced with a presidential deferral of their congressionally-mandated pay increase. Military and civilian government employees' pay increases had been coupled in 1967 and the effective date of increase was made the same in 1974. As a result, AFGE leaders saw a potentially formidable ally in military personnel in their efforts to lobby Congress to override the president's actions. The leadership gave wide distribution to a flyer urging service men and women to write their congressmen to support a measure overriding the pay deferral. The response was tremendous and successful, thus providing the AFGE with the impetus to explore other areas that might prove of mutual benefit to AFGE members and military personnel.

The AFGE had, since the late sixties, represented civilian technicians on duty with the reserves and the National Guard. Their numbers were estimated at approximately 15,000. In addition to the actions of the AFGE leadership, the unionization effort was apparently gaining substantial impetus from the lower ranks of the military services. A 1976 air force study indicated that 36 percent of enlisted airmen would have joined a servicemen's union immediately.* Similarly, a survey of 21,000 soldiers conducted by Citizen Soldier, a GI rights group in New York, indicated that 61 percent of low ranking enlisted personnel favored immediate unionization as of 1977. And by early 1977, Green Beret paratroopers stationed at Fort Devens, Massachusetts, had formed a union organizing committee and, according to the AFGE regional vice-president, were to be chartered as the first military local.

The confluence of rank-and-file sympathy for unionization and the initial efforts at organizing, coupled with the tentative moves of the AFGE, brought about a massive response from those in Congress and the Pentagon who perceived unionization as the ultimate erosion of the traditional notions of military service. Public Law 95-610, signed by President Carter in November 1978, not only made

---

*Only 16 percent of the officers surveyed said that they would join. This figure complements other survey results which suggest that officers are less likely than enlisted personnel to adopt an occupational (rather than an institutional) orientation.

it illegal to join a union but also required the AFGE to cease and desist from organizing reserve and National Guard technicians. This legislation effectively forestalled the full development of the occupational model of military service.

Although the passage of this legislation has ruled out unionization itself as an issue in the foreseeable future, those factors that brought it to the fore will certainly be evident as issues in military personnel management in the 1980s. These issues are likely to become the domain of more traditional service organizations, such as the Association of the U.S. Army and the Fleet Reserve Association, to the degree that these organizations are effective in resolving them. They may become the province of more radical organizations if the issues are left unresolved. Either way, the issues will remain as long as the convergence between civilian and military work continues.

Unionization has not been an issue for the officer corps but the conflict between occupation and calling has been equally troubling there. During and immediately after the Vietnam period, there was a good deal of agonizing self-reappraisal. A principal complaint was about "ticket-punching"—arranging the sequence of one's assignments so as to advance one's career, rather than concentrating on the job to be done at the time. The officer personnel system was— and still is—indicted for putting a premium on ticket-punching. Some outbursts have been expressions of frustration over an apparently unwinnable war and an increasingly untrainable body of recruits. But the more thoughtful reactions reflect the pains of transition from the heroic to the bureaucratic model—realizing that bureaucrats can be heroes too, but it's harder—and the difficulties of preserving and adapting the traditions of a calling to a society that is more and more oriented toward the occupational mode.

The conflict between occupation and calling is reflected in two other areas of military concern: (1) shortfalls in the reserves; and (2) difficulties in retaining career personnel, particularly those in scarce specialties directly competitive with civilian employment. Both problems have been attributed primarily to other causes: the shortfall in reservists to the disappearance of the draft as an incentive to join the reserves; and the retention problem, particularly for specialists, to the military's repeated failure to keep pace with increases in civilian pay scales. But behind both problems lies the underlying difficulty in maintaining noncoercive, noneconomic motivations for military service.

Both the Selected Reserve and the Individual Ready Reserve* have experienced significant shortages of personnel without conscription to motivate individuals to enlist as an alternative to active duty. The absence of the draft has also resulted in a substantial downward trend in the number of reserve personnel in the higher mental categories and with higher educational attainment. The strength of the Selected Reserve declined by over 100,000 during the seven years following the end of conscription, and the supply of individuals in the Individual Ready Reserve declined from over a million and a half to around 400,000 in the same period.

Today, while most navy, marine, and air force units report having nearly 100 percent of the reservists they would need for a crisis mobilization, the army remains some 175,000 troops short. The Individual Ready Reserve, in turn, remains approximately 250,000 short of required levels.

Reservists without prior service who had some college education or a high school diploma or who were in the higher mental categories also have tended to disappear from the makeup of the reserves. The percent of reserve personnel with some college education dropped from over 40 percent in 1970 to around 6 percent in 1979. High school graduates dropped from over 90 percent to approximately 50 percent in the same period, while mental category I and II personnel dropped from over 60 percent to just over 20

---

*The Selected Reserve consists of paid part-time personnel assigned to units of the United States Ready Reserve, which includes all members of the National Guard. The Individual Ready Reserve (IRR) consists of unpaid Ready Reserve personnel not assigned to the Selected Reserve. Together, the IRR and the Selected Reserve comprise the Ready Reserve. Still other individual reservists may be part of the Standby Reserve. They may be ordered to active duty without their consent only during war or formally-declared national emergency. The maintenance of an adequate reserve is considered essential to United States strategy that emphasizes the total force concept (active and reserve components) and depends on mobilization in time of national emergency or war.

percent. As was the case with active forces, these trends have been attenuated by the recruiting upswing of 1981.

What these figures seem to say is that duty in the reserves is seen to be much less a general civic obligation than it once was. One can speculate on how much this change reflects a diminished sense of community in American society—or a sense that the military in its day-to-day activities is less than an integral part of that community. It clearly suggests that the military establishment is not seen as an individual civic responsibility—except as a large part of one's obligation to pay taxes.

Service in the reserves, therefore, is a social problem that can be overcome in part by making it economically attractive in a number of ways. The problem of retaining career personnel—particularly specialists—in the regular forces is primarily an economic problem to begin with.

The military pay problem is a recurring one. After a long period of lagging behind civilian pay in the 1960s, military pay achieved relative parity in the early seventies. In the late sixties, approximately 50,000 military families were eligible for some form of assistance from welfare agencies because they had incomes below the poverty line. This situation was remedied in 1972 through congressionally-mandated pay increases designed to ease the transition to the all-volunteer force. These increases had the effect of eliminating the need for welfare payments for military families and made military pay roughly comparable to that for similar work in the civilian sector. In the last seven years of the seventies, however, the Consumer Price Index rose by over 75 percent, while military compensation increased by just over 50 percent.

There were several reasons why military earnings slipped behind the cost of living. First, the budget process was (and still is) biased in favor of spending more dollars on new and more elaborate weapon systems and fewer dollars on the people who operate those systems. Second, the periodic salary increases for federal employees that were obtained from Congress failed to keep pace with inflation. Third, the way in which comparability with civilian federal employees' wages was calculated tended to operate to the detriment of military personnel. Finally, some of the anticipated military pay increases were reallocated to other forms of benefits rather than to take-home pay. Later, double-digit inflation further exacerbated the problem. By the beginning of the 80s, approximately 110,000 service people and their families were eligible for some form of public assistance.

What all of this meant in concrete terms was that the average soldier, sailor, airman, or marine who was making $9,000 a year fell far short of the $11,546 needed in 1980 to provide a family of four with a "lower" standard of living. Since approximately $20,000 would provide a moderate standard of living, estimates are that about half of all enlisted families were living below the moderate standard and approximately 17 percent below the minimum standard of living. As mentioned above, one must recognize also that the traditional image of the lower ranking enlisted man (or woman) as a bachelor (or spinster) is no longer valid. Between 1964 and 1979, the number of married lower ranking enlisted personnel doubled. In 1981, one-half of the individuals in enlisted grade E-4 (just below the midpoint of the nine enlisted grades) were married, as were nearly one-third of those at the E-3 level. These same individuals in the lower enlisted ranks also were the ones whose basic monthly pay was below the minimum wage. Nearly 600,000 military personnel were earning less than the minimum wage. Even when required to work overtime, military personnel receive no overtime compensation, as was the case for sailors required to work 80 hours in a week on an aircraft carrier.

The low pay received while in the lower ranks was only a prelude to what the average soldier could expect if he or she should be service-oriented enough to reenlist. For example, in 1981, a military police sergeant with six years of service could expect to make $13,200 annually, while his counterpart in the civilian world would receive approximately $20,000. Similarly, a computer programmer working for a civilian company could expect to earn, at $23,000, some $8,500 more per year than a military person doing exactly the same work. The greatest disparity existed in jobs like those filled by navy pilots, who, though earning nearly $28,000 annually after eight years service, could make $20,000 more per year in the civilian world. The result of these enormous disparities was that the navy experienced a shortfall of 35 percent in pilots in 1981, expected to reach 46 percent in 1982. The retention of senior career personnel (those with more than 13 years of service) in all services has dropped steadily and precipitously since 1967 as a result of financial and other problems. Service-wide retention dropped over 13 percent between 1975 and 1979 (although, as stated previously, it rose by some 12,000 persons in 1981). The feeling of career personnel was perhaps summed up best by an air force sergeant who told a *TIME* magazine interviewer that he was "missing the American dream by serving in the American Air Force."

A further attempt to eliminate the lag in military pay was made by an across-the-board increase of 11.7 percent in the fall of 1981,* but these periodic increases cannot match the compensation levels of private industry for jet pilots and computer programmers. Given the sluggishness of the legislative process, it is not likely that the general level of military pay will continue to keep pace with inflation. Staying on in the service will continue to be as much a matter of professional pride and patriotism as of material rewards, even as the hero-warrior image of earlier ages is replaced by the less glamorous but more realistic constabulary image of the nuclear age.

As there are conflicting role models within the military today, so too are there conflicting images of the military role among the public at large. Military virtues have never been enshrined in the United States either as the basis of private virtue or of public morality. Americans have always rejected Prussianism and caudillismo. Until the beginning of the Cold War, the United States military was traditionally ignored by the American people in peacetime and embraced in wartime. The army has been described by former colum-

---

*This increase compared with a 4.8 percent pay hike for comparable civilian government workers. Congress also approved variable military pay increases of 10–17 percent for FY82, with the highest increases allotted to personnel in skilled and experienced levels facing critical shortages, i.e., senior enlisted and mid-level officer ranks. Civilians on the government's payroll were granted a 4.5 percent pay increase for FY82.

nist Joseph Alsop as a desert plant, able to survive long periods without nourishment and then to take full advantage of a sudden downpour, storing up reserves for the next period of drought. With the onset of the Cold War, this situation was reversed. A very large military force became a permanent fact of national life in peacetime while the massive use of force in Vietnam precipitated, over time, a massive negative reaction in the American public—with good and sufficient reason. The military establishment had come to be accepted as a major, permanent institution in American society.

In this context, the military appears more and more like other large institutions in American society. We have already referred to the idea of convergence as developed by military sociologists: The fact is that military work-styles and life-styles in the upper and middle ranks are becoming more and more like those of the corporate executive or white-collar worker. Even the ordinary enlisted man (or woman) leads a life more and more like that of his (or her) civilian counterpart.

The general level of confidence in American institutions has not been very high in recent years. In relative standings, the military does not do badly. For example, a November 1980 Gallup poll showed that 52 percent of those questioned had a great deal of confidence in the military institution; respondents placed it behind only organized religion (66 percent) and banks (60 percent), but ahead of public schools, the United States Supreme Court, newspapers, organized labor, Congress, television, and big business. In a more recent Harris poll, taken in October 1981, the military was rated fourth out of 15 major institutions. Placed ahead of the likes of the White House, organized religion, Congress, law firms, and, of course, big business, the military trailed only medicine, higher education, and the United States Supreme Court.

Yet Americans are more divided about the military than they are about other institutions. Veterans of World War II and veterans of the Vietnam War are often at polar extremes. Young people from deprived backgrounds see military service as a live option or even a unique opportunity to escape poverty and unemployment, while their middle-class counterparts tend to think of military service as a chore that fortunately can be avoided—unless the draft is restored. Americans overwhelmingly believe in the need for increased defense spending, but they are deeply concerned about the size of the defense budget. Some of this ambivalence over an organization that bulks so large in American life is inevitable. If civilian authorities can get a better handle on the governance of the military establishment, as discussed in the next chapter, a good deal of the ambivalence may diminish, but it will not disappear.

The sources of ambivalent feelings about the military run deeper. As citizens of the only nation that has employed nuclear weapons to attack human targets, many in the United States have suffered a recurrent sense of guilt; a guilt that may have been prolonged and accentuated by the Dulles doctrine of massive nuclear retaliation. This burden on the conscience has probably spared the majority of the American people, but for those who had guilt feelings, the Kennedy administration's substitution of the doctrine of flexible response (designed, at that time, to accommodate the use of appropriate force at all levels of conflict) was a matter of considerable moral relief. With the war in Vietnam, however, the attendant revelations of the despoliation of land and villages, and the extent of civilian casualties, the moral liability for the use of United States military power impacted on the consciousness of American society with splintering force. Today many Americans—and at least as many Europeans—have serious questions about an alliance doctrine that contemplates the first use of nuclear weapons under any circumstances. Congress and the administration are divided on the propriety of military assistance—let alone the involvement of American soldiers—in shoring up regimes that brutalize their citizenry. And some Americans raise troublesome moral questions about the reach of United States policy on protecting Middle Eastern oil.

None of these questions is up to the United States military to settle, although military leaders may weigh in with strong opinions. But how the issues are resolved cannot help but affect popular judgments about the military and military service.

What is true for citizens may not be equally true for citizen-soldiers. There is very little discussion of national security policy in recruiting offices, and military training begins by instilling the habit of obedience. That the primary function of today's military is deterrence rather than defense makes the question of what one would fight over less immediate. But one cannot deter what one will not defend, and soldiers cannot help having opinions about what is worth defending.*

Michael Walzer reminds us in *Just and Unjust Wars* that fighting ultimately must be morally justifiable and essentially defensive in nature to insure such justification. Military service is more easily made to look like an occupation in times of peace. It is necessarily more than an occupation in time of war. Everyone needs to know that the wars for which the young may be recruited are ones they can justify within their common value systems. Those who have been encouraged by the 1981 upturn in military recruiting and retention have voiced their concern that improved economic conditions could mean renewed difficulties for the all-volunteer force. It may not be too soon to think about more fundamental justification for military service—and to ask how far our national security policies provide that justification.

---

*A survey of volunteer enlisted personnel undertaken by Moskos and associates in 1975 indicated that 65.3 percent of those interviewed "definitely would volunteer" to fight in an invasion of the United States, while 31.9 percent, 31.7 percent, and 27.6 percent would serve of their own volition if the Middle East-Israel, Western Europe-Germany, or the Far East-Korea respectively were invaded.

VI

Civilian Control

CIVILIAN CONTROL OF the military establishment seems so basic and generally accepted a principle as scarcely to be worth discussing. And yet we are surrounded by distressing evidence that civilian control of today's booming military establishment is a good deal less than a generally prevailing reality.

To take examples only from the decade of the seventies: The chairman of the Joint Chiefs of Staff conducts an espionage operation, including the purloining of secret documents not intended for his eyes, in the offices of the assistant to the president for national security; an air force major general conducts a private bombing campaign against North Vietnam, apparently unsanctioned by his superiors, military or civilian; cost overruns on 45 major weapon systems amount to $31 billion;* and the Pentagon public information and legislative liaison services show no signs of diminishing the size or intensity of their efforts, despite mounting criticism from the outside world.

It is tempting to attribute these out-of-control episodes to the excesses of the power-hungry military. But there is no reason to believe that American military men are more power-hungry than their counterparts in civilian bureaucracies, in or out of government. There are conflicting strains in the training and indoctrination of the career military, so that they are at the same time anxious to receive policy guidance from the civilian side and concerned to

*These particular figures were cited in 1972. Recent (1981) figures indicate that for 17 of the largest army weapon systems alone the cost overrun will approach $27 billion.

protect their professional autonomy—which may, in their view, include judgments about how much is necessary, in men, weapons and dollars, to protect the national security.*

The fact is that the apparatus of civilian control that was developed to implement the original concept of the founding fathers has proved wholly inadequate to control an establishment several orders of magnitude larger and more complex. The problem is not the overweening military but the inadequate civilians who, lacking the means, cannot even test their determination to exercise effective control. The danger in this situation is not that the military may take over the country but that the country is not able to preside over the military.

This is not to suggest that civilians know better than military professionals how to establish a beachhead, or how to design a weapon system. It is to assert that, as an earlier generation discovered that war was too important to be left to the generals, so the current generation has found that planning a military force designed primarily to deter war is also too important to leave entirely either to the generals or to the civilians.

Concern about civilian control of the military has been a continuing, if minor, theme in American political history, beginning with the debates in the constitutional convention. But by and large this concern, as it expressed itself in constitutional and statutory provisions, has been addressed to the supposed danger of the military assuming civilian authority (the Seven Days in May complex), rather than the danger of civilian authority losing effective control of the military qua military. In *The Federalist Papers* (Nos. 22–28), the thrust of Hamilton's argument goes to the need for a standing army in peacetime and to the absence of any internal threat from such a force, given the constitutional protections.

*During the Cuban missile crisis, President Kennedy was disturbed by the inability of the military to look beyond the limited military field. Robert Kennedy recalls his conversation with the president on this subject in his book, *Thirteen Days:* "He said we had to remember that they were trained to fight and wage war—that was their life. Perhaps we would feel even more concerned if they were always opposed to using arms or military means—for if they would not be willing, who would be? But this experience pointed out for us all the importance of civilian direction and control and the importance of raising probing questions to military recommendations."

The three checks on the power of the military provided in the Constitution—the power reserved to Congress to declare war, the two-year limitation on army appropriations, and the specification of a civilian commander-in-chief—were all taken quite seriously by the founding fathers. In fact, they have proved largely irrelevant to the central dilemmas of civilian control in the second half of the twentieth century.

The allocation of the war power to Congress has been a subject of controversy between the legislative and executive branches, not between the civil and the military arm—and even here it has not prevented the executive from conducting two full-scale wars, in Korea and in Indo-China. The last congressional declaration of war was in 1941. These commitments on appropriations (which apply to the army, but not to the navy; the Constitution did not contemplate an air force) have not prevented the Pentagon from making long-range commitments for the development of weapon systems. Congress finds it difficult, if not impossible, to repudiate the two-year limitation. Further, the attorney general has determined that a congressional appropriation of so-called "no-year" funds—funds that need not be expended in the fiscal year for which they are appropriated—is not a violation of the two-year limit.

Lastly, the power of the president as commander-in-chief, vis-à-vis the military establishment, has never been tested in the courts, and the bare constitutional declaration has not perceptibly helped him to resolve the specific problems of control that will be discussed in detail below.

Congress has been scrupulous in supplementing with statutory language the purpose of the constitutional provisions, but it has done so in areas that again have relatively little bearing on the major issues. Congressional actions here look suspiciously like an elaborate effort to lock the side door of the barn, without looking to see whether or not the horses are gone. In the National Security Act of 1947, which first created a (relatively) unified Department of Defense,* Congress specified that the act was not intended "to estab-

---

*The 1947 act actually created the National Military Establishment, which became the Department of Defense in the 1949 amendments to the act.

lish a single Chief of Staff over the Armed Forces nor an overall armed forces general staff." The Joint Chiefs of Staff, created as a body in the 1947 act, were not supplied with a chairman until the 1949 amendments, which were careful to provide that, "While holding office, the Chairman outranks other officers of the Armed Forces. However, he may not exercise military command over the Joint Chiefs of Staff or any of the Armed Forces." In addition, the size of the Joint Staff has been severely limited by statute, and members of the Joint Staff are prohibited from serving as such for more than three years and from being reassigned after less than a three-year interval.

But these highly specific provisions were not designed to protect the civilian authorities from the pressures of centralized military authority. Rather, they were carefully worked out to protect the autonomy of the separate military services, from centralized (military) control. It is perhaps more significant that Congress specifically provided that "the Joint Staff shall not operate or be organized as an overall Armed Forces General Staff and shall have no executive authority." Further legislative provisions stipulate that the individual members of the Joint Chiefs of Staff (who are also the military chiefs of their respective services) are free to present their views to Congress even without being requested to do so—a provision that has been described as "legislated insubordination."

Perhaps the quintessential case of the civilian control issue as it has been dealt with in the past is to be found in the changes that have occurred in the formal role of the individual service secretaries since 1947. In the original National Security Act, the new secretary of defense was given only the powers of general direction, authority, and control, and each of the three service secretaries was authorized to administer his department as an individual executive department; further, all powers and duties relating to the departments not specifically conferred upon the secretary of defense were reserved for the service secretaries. The 1949 amendments increased the power of the Secretary of Defense, while the service secretaries lost their status as heads of executive departments and their membership on the National Security Council. In the 1953 Defense Department reorganization, the service secretaries were inserted into the chain of command for the unified and specified commands precisely so as to improve civilian control. In 1958, they were again removed from the chain of command to the unified commands, but they retained their right of direct access to Congress.

No major statutory changes have occurred since 1958. Major recommendations were made by the 1960 Symington Committee and the 1970 Blue Ribbon Defense Panel; both reports, had they been implemented, would have exerted significant influence on the civilian control question. The Symington report recommended the complete abolition of the military departments, and with them the service secretaries, under secretaries, and assistant secretaries. The Blue Ribbon report, in contrast, advocated greater decentralization, noting that, "No private corporate executive in the world has the managerial responsibility in terms of manpower, budget, variety, or complexity of operations equal to or approaching that resting on the shoulders of a Secretary of a Military Department."

In another area, Congress has been even more explicit, namely in restricting the authority of the director or the deputy director of the Central Intelligence Agency by means of the statutory provision that one, but not both, of these positions may "be occupied simultaneously by commissioned officers of the Armed Services, whether in active or retired service."

Section 102 of the National Security Act provides that, if a commissioned officer is appointed to one of those posts, "he shall be subject to no supervision, control, restriction, or prohibition (military or otherwise) other than would be operative with respect to him if he were a civilian in no way connected with the Department of the Army, the Department of the Navy, the Department of the Air Force, or the Armed Services or any component thereof."* But

---

*Congress provided additional restrictions by specifying that a military officer named director or deputy director "shall not possess or exercise any supervision, control, powers, or functions (other than as Director or Deputy Director) with respect to the Armed Services or any component thereof, the Department of the Army, the Department of the Navy, the Department of the Air Force, or any branch, bureau, unit, or division thereof, or with respect to any of the personnel (military or otherwise) of any of the foregoing."

these minutely specific provisions do not affect the representation of the military departments on the National Foreign Intelligence Board, which formulates the official positions of the intelligence community.

Thus, we can see that constitutional and statutory restrictions on military power have been addressed to the specter of the man on horseback rather than to the need for civilian authority to set the guidelines by which the military goes about its own business—now the biggest business in the United States.

Before turning to the question of new means to improve civilian control of the military establishment, we need to consider briefly the question of the adequacy of civilian will and its determination to do so. Civilian will and determination are significantly limited by the military-industrial-labor-congressional complex alluded to above.

The very nature of the existing complex tends to sap the will and determination of the politicians in the legislative branch, and even the president himself, to exercise effective control. The political pressures generated by the far-flung constituencies of the military establishment are increased by the politician's reasonable fear that if he is reponsible for denying a military request, and the national security is then impaired—or believed to be impaired—he will be blamed.

The pressures resulting from economic considerations are likely to bear more heavily on the legislative branch than on the executive branch because the economic consequences of changes in force structure are felt on local employment rolls; members of the legislative branch are generally more responsive to local pressures than the executive branch. Pressures resulting from concerns about national security are likely to weigh more heavily on the president, since he is the individual ultimately reponsible. The distinction between these two branches of government is reinforced by the fact that Congress exerts its control over the military primarily by refusing to appropriate money, while the executive branch may take substantive initiatives on changing force structure and strategy. Traditionally Congress has even been reluctant to make specific budget cuts because of the extraordinary amount of technical detail involved; they have generally preferred to make overall budget adjustments or to introduce procedural requirements.* Congressional alliances with the military hierarchy are a well-known fact of life, particularly links between senior members of the relevant congressional committees and senior military officers. On the other hand, Congress as a body is probably less reluctant than the executive branch to take on the military directly on broad issues since Congress does not operate under the same constraints of having to govern the military establishment with and through the military itself.

As the will to exercise civilian control is a variable quantity, so the nature of the control that is appropriate and feasible, even under ideal circumstances, also varies considerably. Whatever civilian control has meant over the past 200 years, it can mean a number of quite different things today. Its scope extends from control of research and development decisions affecting future weapon systems to control of actual military operations—and the degree of civilian control varies widely across this broad spectrum of activities. Part of this variation is attributable to the appropriateness of varying degrees of civilian control, and part of it is due to the relative effectiveness of the control that is actually exercised. Actions of field commanders cannot and clearly should not be controlled as closely as actions of deskbound officers; but rules of engagement, which may be critical to the political conduct of a war, can be more or less effective depending on how they are drafted and on how consistent they are with overall foreign policy objectives.

Civilian control has different meanings in different contexts. It can mean control over aberrations from established policy or over determination of new policy directions. It can mean a general curb on the expansionist tendencies of some military, or it can mean a strengthening of some military activities, undervalued by the military itself, partially or wholly at the expense of others. And, as

---

*As congressional staffs expand in size, this tendency appears to be changing somewhat as, more and more, Congress attempts to influence spending on specific weapon systems. See chapter 4 above.

indicated above, civilian control can be exercised primarily by Congress, by the White House (and its appendages, the National Security Council staff and the Office of Management and Budget), or by civilians within the Defense Department itself.

No matter who exercises control within the civil establishment, that exercise will be either irrelevant or obstructionist unless the civilian authority has a clear view of the national and international framework within which that control will be exercised and of the purposes it wishes to accomplish. Otherwise the issue of civilian control will become a zero-sum game between the military and the civilians, and among the various factions on both sides. Power for its own sake is not worth the candle. Some of the purposes for which civilians may wish to exercise control appear in chapter 2 above; others are spelled out in chapter 9 below. But so long as the reason for the exercise is not to keep the military out of civilian affairs, but rather to assure civilian input into military affairs, civilians must choose their goals and decide on their priorities. It is the failure, or the shortfall, in this pursuit that has muddled national security policy since Vietnam—and perhaps earlier.

The simplest way to increase civilian control across the entire spectrum of military activity—by reducing the size of the military establishment—is not a realistic prospect for the foreseeable future. And a reduction of 20, 30 or even 40 percent in the military budget, beyond anything that was proposed in the budgetary debates of the seventies, would still leave an institution uniquely large and complex. Further, the process of cutting back on budget and force structure would tend to freeze attitudes and heighten intrainstitutional jealousies. The reforms of the early sixties were made easier because they were carried out in the context of an expanding defense budget.

Since the simple way is not open, even if it were reasonable, it is necessary to break down the problem into its component parts, and to look at ways to increase civilian control of (1) overall budgets; (2) research and development; (3) force structure; (4) contingency planning; and (5) actual military operations—a catalog arranged in ascending order of difficulty, the order in which we shall address it.

If this is a novel way of approaching the problem of civilian control, it is dictated by the size and shape of the problem today.

Civilian control of the overall military budget has been a matter of close attention to detail on the part of the civilian authorities in the Pentagon—and of resolute avoidance of detail by the civilian authorities in the White House. The Kennedy administration used to take pride in its ability to determine the budget necessary to protect the security of the United States without regard to arbitrary or predetermined budget ceilings. No one asked at the time if similar instructions had been given to the secretary of health, education, and welfare, or to the administrator of the Housing and Home Finance Agency—or, if they had, how the government would have been able to put together a manageable budget. In fact, President Kennedy was saved from the consequences of his instructions only by the extraordinary ability of his secretary of defense to focus on questions of details as they arose in literally hundreds of "subject issue papers" during budget season. Even then, there had to be some tacit understanding between the secretary and the president that the budget would be more (or rather less) than the sum of its parts. Somewhere in their discussions about specific budget issues, a budget ceiling was arrived at.

What drives the size of the peacetime military budget is as much pressure on the total federal budget from nonmilitary spenders as the actual military needs of the country. Because the decision is so quintessentially political—a matter of choice in which the components cannot possibly be quantified, and the pressures must be balanced largely by intuition—civilian control is less difficult to achieve here than in any other area. Most military pressures for specific weapon systems can be accommodated within almost any given budget ceiling by making appropriate adjustments elsewhere in the budget.

To understand the issue better, it may be useful to look at the process by which the budget ceiling is developed. Before the advent of the McNamara budgeting reforms in 1961, the allocation of the military budget was largely determined by bargaining among and within the military departments. Bargaining took place after a ceiling had been fixed by the president with the advice of the secretary of the treasury. McNamara introduced three major reforms: (1) organization of the budget by functional categories (e.g., strategic forces; general purpose forces; research and development, test and evaluation) cutting across service lines; (2) multiyear programming, projecting the total costs of forces and weapon systems, including systems still under development, at least five years ahead; and (3) intensive review and examination of budget submissions from the military departments by civilian staff in the Office of the Secretary of Defense. The civilian staff would address questions of the military justification for particular weapon systems or system characteristics as well as traditionally "civilian" questions of cost and technical feasibility.

Since the McNamara system did not involve any predetermined budget ceilings, it tended to result either in increased budgets or in major eleventh-hour cuts as the secretary and his staff reviewed the final total program costs for budget purposes. The changes in the system introduced by Melvin Laird (secretary of defense, 1969–1973) provided for initial dollar targets for the military departments, based on preliminary and general planning documents (sometimes characterized as "wish lists") so that the final budget decision by the secretary did not come as quite as much of a shock to the military departments and their bureaucracies. These preliminary budgetary determinations also came to involve review by an interdepartmental group including the secretary of state and the assistant to the president for national security affairs. At the same time, Secretary Laird reduced the role of his civilian staff in proposing program directions and left the initiative more to military planning staffs in the military departments.

Given this sequence of decision making, the major substantive issues in the budget—whether or not to build another nuclear carrier; whether or not to add (or subtract) an army division or a Marine Corps brigade; whether or not to proceed with MIRVing the Minuteman missiles—are settled privately between the Pentagon and the White House, although there may be some judicious leaking to the press by both sides while discussions are in progress. The president will then fix the size of the military budget. He will undoubt-edly be influenced in his judgment by decisions arrived at previ-ously on specific issues. If, for example, he has agreed to buy a new carrier, he cannot preserve a level budget (even in constant dollars) unless he agrees to a cutback in some other part of the force struc-ture. But the decision is the president's, in the last analysis. Even if Congress tries to add a major Pentagon proposal earlier rejected by the White House, it is unlikely to succeed. Defense legislation is not like a rivers and harbors bill. The White House is on stronger ground in refusing to spend money appropriated for defense than it is for domestic appropriations.*

The public issue is likely to be the extent of the increase or reduction from the previous year's budget. Whether or not what is proposed by the administration amounts to an increase may, how-ever, be a controversial question. When the FY 1975 budget re-quest for the Department of Defense was sent up to Capitol Hill, it was accompanied by an FY 1974 supplemental budget request for $6.2 billion. The supplemental included $3.4 billion for pay in-creases authorized by Congress after the original 1974 budget was adopted and $2.8 billion to cover fuel price increases and extra costs incurred in supplying equipment to Israel during the Middle East War. But, it also included $2.1 billion to augment inventories and to buy new weapons and equipment. Except in an emergency, this kind of request is not normally included in a supplemental, but, when it is, it has the effect of increasing the current year's budget and reducing the succeeding year's budget; thus, if the succeeding year's request is higher than the request for the current year, the politically critical differential is reduced.

If, on the other hand, the civilian authorities wish to propose a substantial decrease, there is no simple device to minimize its polit-ical impact and it may present serious political problems to the president and to Congress. Congress is probably in a better position to make across-the-board cuts in the budget than to quarrel with the massive expertise of the military on specific cuts. Also, an across-

*If, on the other hand, the president contemplates a major increase in the military budget at a time of general fiscal stringency, as President Reagan has done, he may find the budget a major political issue between himself and Congress.

the-board cut cannot be identified as affecting the constituents of a particular congressman or senator, while a specific cut can be traced down to the particular defense contracts and subcontracts it will affect. But the knowledge that even an across-the-board cut may have a significant impact on the economy of his district or state cannot help but have some deterrent effect on a legislator and makes it more difficult for Congress than for the president to take the budget-cutting initiative.

In order to maintain effective civilian control over the military budget as a whole without losing control of its basic components, the civilian authorities must involve themselves deeply in the second and third issues, control of research and development, and control of force structure. The principal difficulty civilian authorities encounter is in mastering the technical complexities involved in choosing among weapon systems or in deciding as to the need for a new version of an existing system. The basic problem of lack of expertise is exacerbated by five factors:

First, the decision-making process for new weapon systems is almost insidiously gradual.* Typically, it may take 10–12 years to field a new system. A new weapons concept is explored because it looks promising, because it costs relatively little to explore, because it is thought to be important not to fall behind in the research and development race, and even because exploring new ideas is a necessary activity in order to hold together a first-rate research and development team. When the new idea turns out to be worth consideration, it is only natural to put a little more money into it, and

---

*The present systems acquisition process has evolved generally from the process instituted by Deputy Secretary of Defense David Packard in 1969. Today, the process consists of four phases of activity, each of which is preceded by a key milestone decision. Milestone 0 involves the establishment of a need for the system, without any commitment to what kind of system would best meet the need. Passing this milestone results in the authorization to proceed into phase 0—concept exploration—which includes solicitation, evaluation, and competitive exploration of alternative system concepts. Milestone 1 involves the selection of alternatives and the authorization to proceed into phase 1—demonstration and validation. Milestone 2 involves the selection of an alternative (or alternatives) and the authorization to proceed into phase 2—full-scale development—which includes limited production for operational test and evaluation. Approval to proceed with phase 2 means an intent to deploy the system. Finally, milestone 3 is the authorization to proceed into phase 3—production and deployment. It has been estimated that over 70 percent of total life cycle costs for any weapon system are *committed* by milestone 1.

then yet more. Turning back on the road to full weapon development becomes more and more difficult as the investment in the idea grows. Many ideas are abandoned at various stages because they become impractical, although even as impractical an idea as the nuclear-powered airplane was pursued for years before the decision was made to drop it. But to abandon a really promising new weapon system, only because it isn't really needed—or because its job can be done more cheaply—is the most difficult decision of all.

Second, neither the secretary of defense, nor the president, nor Congress separately, nor all of them together, can marshal the staff of experts to match the staffs of the military departments, even if those staffs were reduced to the absolute minimum necessary to carry on enough research and development to keep up with our major foreign antagonists.

Third, at least some of the essential facts necessary to exercise reasoned judgment about research and development alternatives are classified. While civilians can and do penetrate the veil of secrecy when they have established a need-to-know, the very fact that they must establish such a need makes it difficult to ascertain in advance what information may be missing. How does the civilian decision maker or his staff person come to know what it is that he needs to know in order to make informed decisions on research and development issues?

Fourth, in addition to the impact of the classification process properly employed, there is the inevitable use of the classification system to hide adverse facts that the proponents of a particular weapon system would prefer not to disclose. Curbing the abuse of the system is extraordinarily difficult, and even major reforms are unlikely to prevent the kind of abuse involved here. There is no automatic way in which a military research and development staff can effectively be required to disclose, even to properly authorized outsiders, all its fears and concerns about a particular development. *

*Executive Order 12356 on national security information, issued by President Reagan early in 1982, seems to exacerbate the problem by eliminating the "balancing" provision previously in effect. This provision required the classifying official to balance the public interest in disclosure against the national security interest in secrecy.

Lastly, there is an understandable reluctance on the part of civilian outsiders to make adverse judgments on complex issues of military research and development; a wrong decision against a weapon system could, just possibly, mean defeat for the United States in a future conflict, while a wrong favorable decision would only mean unnecessary dollars for defense. That those dollars might be spent for a more needed military purpose, within budgetary limits, is a logical argument, but not an emotionally persuasive one when a particular weapon is at issue.

The more troublesome area for civilian control of weapons development is the nonnuclear one, despite the fact that the most publicized debates concern the choice of nuclear weapons: the MX, the B-1, a new generation of nuclear attack submarines. Nuclear weapons development options are fairly clearly identified with specific strategic policies. The issues here are complicated. The military can usually be counted upon to weigh in on the side of new and more complex systems (unless the system is caught in the cross fire of interservice rivalry).* But strategic policy decisions are ultimately made by civilians, and the weapons development decisions follow. On the nonnuclear side, however, the characteristics of the new main battle tank or the loiter time for a new strike aircraft are not as clearly identified with particular strategies, and the military pressure to develop weapons to meet all possible requirements is greater. The extraordinary difficulties experienced with the F-111, originally designed to be the first two-service aircraft, were probably not in the basic, civilian-inspired concept but in the conflicting requirements imposed by the two services. The two sets of requirements were never translated into alternative strategies and evaluated by civilian authorities.

Perhaps the best preparation for a civilian manager who must deal with decisions about developing new weapon systems is an intensive course in the classical schoolboy art of précis writing. He

---

*The first of these interservice battles developed in 1949 when the air force sought approval of the B-36 jet bomber as the centerpiece for the nation's emerging deterrence strategy. The $15 billion defense budget could not allow both the B-36 and the fleet of new supercarriers sought by the navy as a supplementary means of getting nuclear bombers to distant foreign targets. The air force got its B-36, but the navy's supercarriers were denied.

must be able to break through enormous technological complexities to the basic issue of military usefulness, and he must be able to translate the most elaborate cost-benefit analysis into layman's terms. If he fails, his superiors in the Pentagon and the White House will either have to spend more for defense than is reasonable in light of other national needs (including other elements of national security), or they will have to buy a force that is overequipped in some ways but inadequate in other ways—most likely in sheer numbers—to meet national defense requirements.

Even more difficult for civilians to cope with are issues of force structure—particularly nonnuclear force structure. Again, the question of the mix of manned strategic bombers and land-based and sea-based nuclear missiles is relatively uncomplicated as compared with the question of what kind of divisional structure is appropriate to the kinds of nonnuclear conflict that the United States might become involved in—or what should be the shape and size of the Rapid Deployment Force. These nonnuclear force structure issues are becoming increasingly critical because they involve much more manpower (as well as a larger share of weapons and equipment costs); and, without the draft, the availability of volunteer manpower on any reasonable pay scale may be the limiting factor on the size of the military establishment, as discussed in chapter 6 above. Also, force structure issues have the most direct political implications, as, for example, in the question of substituting a militia-type maneuver-oriented defense structure for many of the standing armies in the defense of Western Europe.

Force structure issues are seen as even more "military" than issues of weapons development. They are not as technologically complex, but they involve more imponderables—readiness, combat-to-support ratios, heavy versus light configurations, to name but a few—and 'there are fewer civilian analysts competent to deal with them. It is more difficult, therefore, for civilian authorities to get a handle on these issues. Traditionally, force structures were taken for granted; decisions about them were entirely within the province of the military. If they are to be brought into the arena of civilian reexamination and ultimate decision making, the first requirements will be to develop a body of civilian expertise sufficient to critique military practice and to explore the implicit major premises of military proposals.

If force structure issues are more "military" than issues of weapons development, then contingency planning and actual military operations are closest to the bone in the structure of military professionalism and military autonomy. It was not so long ago that the secretary of war did not have access to the war plans of his general staff. The war plans for nuclear conflict are now a subject of intense political discussion, while nonnuclear contingency plans, particularly in Western Europe, precipitate serious disagreements among allies. The principal device for civilian control of military operations is the promulgation of rules of engagement—rules governing the circumstances under which forces can be committed to military action.

In direct confrontations between nuclear powers, not only the rules of engagement but the details of their application as well are determined by civilian authorities, as they were in the Cuban missile crisis or the several Berlin crises of the fifties and sixties. Since the danger of escalation is so great, and the application of nuclear force is recognized as involving the life or death of civilizations, the closest control of such military operations by civilian authorities, often working directly out of the White House, is assumed— although not always worked out in practice.* The popular notion of a "button" that someone—presumably the president—can push, assumes away both the complexities of the decision-making process and the dangers of independent actions in the field. Jeremy Stone's suggestion of a (very limited) consultative process highlights the consequences of allowing one human being to invoke Armageddon.

The principal limiting factor on civilian control is the refinement of so-called command and control technology, which enables headquarters in Washington to maintain rapid communica-

---

*During one crisis over allied access to Berlin by road through East Germany, an interagency committee considered whether the convoy commander should acquiesce to the Soviet demand to lower the tailgates of the trucks, or should only allow Soviet soldiers to peer into the trucks over the tailgates.

tion with the field, even under quite adverse conditions, and in some cases to be able to control physically the operation of nuclear weapons.* The overseas deployment of large numbers of tactical nuclear weapons further complicates the problem. Clearly, the danger, however limited, of an unauthorized or inadvertent nuclear exchange needs to be seen in the light of the potential disaster that would result. The level of the danger at any time is generally in inverse ratio to the degree of concern at the highest levels of government—in the National Security Council staff and the White House Oval Office.

The war in Southeast Asia was a tragedy initiated by civilians and persisted in by civilian decision. More effective civilian control of the military could not have redeemed it. But as a case study in civil-military relations, it confirms the proposition that civilian control of military operations in a modern guerrilla war (or even a semiguerrilla war) cannot be maintained only by issuing orders from civilian command. The inadequacy of civilian command is even more evident when the military problem is one of sporadic low-level violence and terrorist threats. The problems are too complex, the interplay of political, psychological, and military considerations too subtle. Senior military commanders need to understand and share the objectives of their civilian superiors.

At the same time, it must be recognized that, at some point, civilian control may become unjustified interference in professional military functions. Nowhere was this dichotomy more clearly demonstrated than in the abortive attempt by the United States on April 24, 1981 to rescue American hostages held captive in Iran. Civilians seemingly called even the most technical planning and execution shots. When three of the eight helicopters involved in the operation went down because of equipment failure, the mission was ordered aborted. Another helicopter then crashed, killing eight American soldiers. It is impossible to predict how the mission would have fared had the military been unfettered within its own domain of

*Recent controversy over targeting policy for Soviet command and control centers (and concern about Soviet policy for United States command and control) raises the question whether any controls can be exercised once these centers are destroyed.

expertise. Nonetheless, it is clear that civilian takeover of purely operational functions is very likely to produce acute dysfunctions. The Iranian rescue fiasco might be contrasted with President Kennedy's detailed and delicate handling of the boarding of a Soviet freighter on the high seas during the Cuban missile crisis. In a crisis situation, the quality rather than the quantity of civilian intervention is critical.

Achieving this kind of understanding is perhaps the most difficult task for the civilian management of the military. Every bureaucracy suffers from some form of professional distortion. President Kennedy's frustration with the State Department bureaucracy during his administration is well known: he characterized it as "a bowl of jelly." Other characterizations might be more appropriate for the Pentagon, but they can be equally frustrating—or even frightening. The military men are trained to achieve at any cost the objective toward which it is directed. The senior author recalls the colonel who came into his office in the Pentagon one day for an assignment, and announced: "Just tell me what to do, sir, and then I'm all arms and legs." Military personnel are not as well prepared to examine the incidental costs of achieving those objectives. They work in an institution that is remarkably effective at avoiding the gross phenomena of disorder, and they are therefore inclined to minimize the likelihood that any of their actions will result in gross disorder. And military personnel are encouraged to compete fiercely for advancement within their own system, so that they are somewhat less responsive to the judgments of outside observers.

Samuel Huntington, writing in the late 1950s, first drew the distinction between "objective control" and "subjective control" of the military. The former encouraged military professionalism and concomitant respect for civilian authority, while the latter instilled in the military the substantive social philosophy of the civilian authorities. Huntington, in his influential book, *The Soldier and the State*, expressed a preference for objective control. But he was writing in an earlier age.

The phenomenon of "convergence," discussed in chapter 6 above, makes objective control of the military more difficult. It does not necessarily make subjective control any simpler. One must avoid the easy but dangerous assumption that greater exposure to the processes of civilian life automatically carries with it greater understanding of the nonmilitary aspects of United States foreign policy. A military man or woman, like his or her civilian counterpart, can be quite at home in the institutions of civil society, and at the same time can accept uncritically the most superficial assessments of and assumptions about political realities that are common in that society.

Enhancing civilian control of actual military operations, then, is to some extent a matter of education, and the educational process is a lifelong one. The process extends from efforts to broaden the curriculum of the military academies and to increase the proportion of officers coming into the services through ROTC programs to efforts to provide greater in-service educational opportunities in high-quality civilian institutions, particularly at the graduate level. It includes the creation of satisfactory career lines for officers specializing in politico-military relations and even the possibility of lateral entry for civilian specialists into the officer ranks.

One of the ways to educate the military is to involve it more extensively in formulating national security policy, not as a by-product of Pentagon decisions about weapon systems and force structures, but explicitly, as valued professional advisor on the military implications of alternative foreign policy choices. So many internal Pentagon decisions do have foreign policy implications which the military ought to bring to the attention of its civilian colleagues, but it cannot be ordered to do so. It may, however, be persuaded by the same spirit of comity that reputedly led Secretaries Acheson and Marshall to agree, as quoted earlier. And if military people are taken more into the confidence of their civilian opposite numbers, they are less likely to set up informal intelligence gathering networks that can proliferate, with typically single-minded military efficiency, into full-fledged espionage systems.

The responsibility for change must not be placed solely on the military. Just as civilians not infrequently bemoan the "military mind," so too do military professionals cringe at the prospect of taking orders from "amateurish" defense intellectuals (or pseudoin-tellectuals). Few military professionals deny the validity of civilian control. In the words of the 1978 *Report of the Secretary of Defense on the National Military Command Structure* (the Steadman Report), "we find that the concept of civilian control over the military is unquestioned throughout the Department. It is a non-issue." In this sense, the acculturation process has shown itself to be remarkably effective. But, these same individuals, like most of us, want to know that they are in the hands of competent and understanding leaders. It becomes no less important, therefore, for civilians who occupy positions of authority within the military establishment to understand military values than for their military counterparts to appreciate civilian concerns.

Mutual trust and confidence are in no sense sufficient conditions for effective civilian control of the full range of military activities. But trust is clearly a necessary condition. Civilian control requires presidential will and determination, congressional willingness not to insist on protecting local economic interests, and extraordinary expertise and energy in the secretary of defense and his staff. It depends also on changes in attitudes toward the military in the government and in the country: less concern about the prospects for irresponsible military interference in civilian affairs must be replaced by more concern about the need for responsible civilian involvement in military affairs.

VII

Alternate Futures

IT USED TO BE that the most probable future for any institution was not likely to differ very much from its present or its immediate past. On this basis, the United States military establishment is not likely to look drastically different ten or even twenty years ahead from the United States military establishment today—always excepting the possibility of a major war.

It is still true that the technological innovations for the next ten years are already on the drawing boards; the military-age population for the next fifteen years is already out of diapers; and the flexibility of the peacetime domestic political system in shifting the allocation of resources either between the public and private sectors or within the public sector remains quite limited—although the Reagan administration is testing the limits of that flexibility, as will be noted below.

But we are also aware of major discontinuities in our society, so much so that a new priestly class of futurists has emerged as prophets of change. They have identified the new era as postindustrial and proclaimed the shift to an information society, organized around the computer memory chip. This revolution is announced to be as significant as the shifts from a hunter-gatherer society to an agricultural and then to an industrial society. The futurists have noted the increasing interdependence within the global village and the trends toward a world wide economy and a global shopping center. At the same time we are warned that, absent more effective planning, we face critical energy and raw materials shortages and an exploding world population problem. The have-not nations, which make up a majority of the present population of the planet, are demanding a major redistribution of resources, while the 1980 Global 2000 Report, prepared for the Carter administration, predicts that, without remedial measures, the gap between rich and poor countries will actually widen over the next generation.

In the industrial societies of the West, no cure has yet been found for the newly endemic problems of stagflation, while the increasing vulnerability of these societies to low-level violence and terrorism is a growing cause for concern. The burdens of sustaining an aging population, a greater proportion of whom are beyond the traditional retirement age and are consuming increasingly ingenious and expensive medical care, weigh on the more advanced societies—as they do on military retirement costs which already take up more than 8 percent of military spending.

There are a number of possible scenarios for the United States military establishment in this uncertain future. Four of these scenarios will be sketched in here: a beleaguered armed camp; a severely constrained military budget; major budget increases as proposed by President Reagan; and a new emphasis on military reform—which, in turn, could go off in several directions.

The first scenario—perhaps least likely, if most alarming—is for the United States to become a garrison state, a fortress America. Growing neutralism in Europe, as well as in the Third World, and a failure of American diplomacy to counter these trends, could lead to withdrawal of United States forces stationed in Europe and perhaps in some other parts of the world. Whether the result would be expansion of the Soviet empire or the further spread of neutralism, it would compound the sense of isolation that gave rise to the policy of withdrawal.

Bringing American forces home would not save money, at least initially, although it might reduce foreign exchange costs to the extent that they are not borne by NATO allies. It would make the peacetime military establishment more visible on the domestic and on the local scene. Paradoxically, greater visibility might coincide eventually with a shrinking defense budget. It would be difficult, without major overseas commitments, to justify a large, two-ocean navy, although it would no doubt still be argued that, in the event of all-out war, carrier task forces would be needed to carry the war to the enemy while attack submarines fended off ocean-borne aggression. Without their peacetime missions in Western Europe and in Korea, the army and air force would almost inevitably shrink. And even a further enlarged strategic nuclear force could not soak up all these savings; one of the most dangerous aspects of nuclear weapons is their relative cheapness.

The expansion of nuclear forces could take various forms. It might begin with antiballistic missile deployment around land-based ICBM installations, probably precipitating abrogation of the ABM treaty (discussed in chapter 8 below). A kind of national claustrophobia might then lead to a revival of the concept of area defense against nuclear attack by deploying ABMs around population centers; these deployments, in turn, would lead to still greater uncertainties about the effectiveness of our second-strike capability as the Soviets emulated our ABM deployments. Matching Soviet deployment would tend to increase the arsenals of offensive nuclear weapons on both sides and the attractiveness of a preemptive strike.

Resurgent isolationism could revive enthusiasm for massive civil defense, and obliterate the lessons of the Kennedy administration's abortive experiment with an expanded civil defense program, which collapsed in a flurry of debate over the morality of shooting one's neighbors when they sought to share one's private fallout shelter. There are renewed rumblings from the Reagan administration about massive evacuation plans for a nuclear crisis, despite overwhelming evidence that the most elaborate evacuation scheme, American or Soviet, could be neutralized by relatively minor shifts in the targeting plan for strategic nuclear weapons. The strongly negative reactions of the grass roots to current evacuation plans make this development unlikely. It seems even more unlikely that the United States would attempt to emulate the Swiss and the

Swedes, with their very different vulnerabilities, by honeycombing urban areas with deep underground shelters.

Without effective alliances, foreign trade would be stifled because it would be thought that the country could not afford to rely on sources of supply outside the continental United States. Economic autarky would reduce the gross national product and lower the general standard of living with consequent domestic political unrest but with even greater disturbance to the economies of the less-developed countries. They would be effectively cut off from the trade and assistance necessary to improve their economies and to control the rising tide of population growth. The United States could, with some severe transitional pains, adapt to the cutoff of strategic materials and of fossil fuel sources from outside the Western Hemisphere. The consequences of a collapse of international trade for Europe, Japan, and particularly for the Third World would be no less than disastrous.

Rising unemployment in the United States would make restoration of the draft unnecessary, particularly with a smaller force structure. But compulsory universal service, both military and (with limited military manpower requirements) civilian, might be instituted because of the ravaging social effects of dangerously high unemployment figures for young people.

Perhaps the most alarming consequence of this scenario would be the horizontal proliferation of nuclear weapons to other countries and, as has been suggested above, inevitably to political factions within some countries and eventually to private terrorist groups. The reduction of superpower influence and the probable disappearance of even rudimentary international institutions would encourage this development. In sum, this scenario could be characterized as a nuclear holocaust waiting to happen.

The second scenario, at the opposite—and almost equally unlikely—end of the spectrum, envisages a significant decrease in military spending below the FY 1983 Reagan budget increment. Such a decrease could presumably result from greatly increased pressures on the discretionary portion of the overall federal budget, currently estimated at some 24 percent, of which 14 percent is military and 10 percent civilian spending. These pressures could in turn be a by-product of continuing inflation, recurring recessions, social unrest, and the inability of the system to make internal adjustments.

The dangers of unraveling alliance budget commitments are apparent in the absence of effective new arms control agreements. Consequent reductions in force structure could, but would not necessarily, result in some scaling back of international political commitments.

Budgetary constraints might lead to consideration of less expensive, less advanced, and less complex military technologies. But since the United States military seems to have particular difficulty in constraining the complexity of new weapon systems, a more probable route would be to cut back actual numbers of missiles, planes, tanks, and ships to be purchased. There might well be a political tug-of-war between those who felt that because the United States could not give up its traditional European commitments, its friends in Japan and Korea would have to be left to fend for themselves. Conflict would also occur between those who fervently believed that westward the path of empire still stretched and that the greater political and economic opportunities were to be found in what is still called the Far East. One can even imagine a situation in which the rising tide of economic and political change in Latin America would focus the attention of the United States military establishment on that area—however inappropriate and even self-defeating the military instrument might prove to be.

How rationally the military establishment would react to substantial budget cuts (or increasingly constraining budget limitations) is a question that has no obvious answers. One can only offer the general observation that it is easier to improve the efficiency of an expanding organization than of a contracting one, since in a period of contraction each unit and subunit tends to draw its wagons into a circle and take up firing positions against attacking budget-cutters.

There is still a possibility that imaginative military and civilian leadership could take advantage of the situation to produce a leaner, tougher military establishment, perhaps even one that relied much more heavily on reserves, intensively and realistically trained, to fill out the skeleton units held together by a highly professional cadre. This possibility is explored in greater detail in connection with the fourth scenario below.

The third scenario is based on the current budget projections of the Reagan administration, and contemplates major increases both in commitments ("Total Obligational Authority", or TOA) and in outlays, as far ahead as the projections extend. From the $196 billion in TOA and $180 billion in outlays originally proposed in the Carter administration budget for the fiscal year ending September 30, 1982 (and representing a significant increase over the FY 1981 budget), the Reagan administration increased the FY 1982 spending figure only to $181 billion and the TOA figure only to $200 billion (although the original Reagan revision of the Carter budget was for $222 billion). The cumulative effects appear in the later years; projections for fiscal 1983 through 1987 appear in table 7.1. These figures anticipate a spending total of $1.6 trillion over the five-year period, and, if approved by Congress, will result in an increase in the proportion of the federal budget devoted to the military from 28.5 percent to 37 percent; and in the percentage of gross national product so committed from 6.3 percent to 7.4 percent. At this writing, it appears that these figures will be cut by Congress, but probably not by a large enough figure to alter the general shape of this scenario.

Some observers believe that even if these increases are approved by Congress, the military establishment simply cannot absorb them at the rate proposed. Some argue that many of the increases will be lost in price hikes for major equipment well beyond the general inflation rate. Since the capacity of existing defense producers is limited, cutting into the defense market is quite difficult, as discussed in chapter 4 above. At the same time, these proposals fall far short of the "wish list" compiled by the Joint Chiefs of Staff for the forces and the equipment they believe would be needed to meet the administration's national security objectives. Their list apparently includes nine more carrier task forces, fourteen

TABLE 7.1   Administration Defense Budget Projections,
Fiscal 1983–1987 *(Billions of Dollars)*

	TOA			Outlays		
Fiscal Year	Total, Current Prices	Total, Constant (FY 83) Prices*	% Change from Previous Year in Constant Dollars	Total Current Prices	Total Constant (FY 83) Prices*	% Change from Previous Year in Constant Dollars
1983	258.0	258.0	13.2	215.9	215.9	10.5
1984	285.5	269.8	4.6	247.0	233.2	8.0
1985	331.7	297.8	10.4	285.5	255.6	9.6
1986	367.6	314.0	5.4	324.0	276.0	8.0
1987	400.8	325.9	3.8	356.0	288.7	4.6

*The following inflation rates were employed in calculating the constant prices: FY 82 to 83, 6.1%; FY 83 to 84, 4.9%; FY 84 to 85, 4.6%; FY 85 to 86, 4.5%; FY 86 to 87, 4.4%.

more air force fighter wings, and nine more army divisions, as compared with one more carrier task force, three fighter wings, and one army division in the administration's budget proposals through fiscal 1987.

The Reagan administration's proposed increases involve some expansion of the numbers of men and women in uniform. But, as discussed in chapter 5 above, that expansion is constrained by the limited ability to attract volunteers from a declining 18-year-old age cohort, and at this writing the political possibilities for a reinstituted peacetime draft seem to be minimal.

The bulk of the new spending is likely to go for machines rather than for people. The MX missile, the B-1 and Stealth strategic bombers, and the additional Trident submarines are large cost items, and the new tanks, ships, and planes for the nonnuclear forces are even larger ones. The bias of the system toward developing and procuring new and more complex weapon systems, rather than spending scarce dollars on training, maintenance, and general readiness is discussed in chapter 4 above, as are the impacts of additional spending on the general economy, on energy supplies, and on the environment. Whether Congress will be willing to authorize and appropriate funds for the full expansion called for in this scenario remains at this writing very much an open question.

Assuming, as seems likely, that there are some cuts from the level proposed by the Reagan administration, these cuts (or rather these smaller increases) are likely to come in the areas of training and maintenance, just because these are the areas where current spending can be most readily reduced. The initial costs of planning and developing a new system are a relatively small proportion of the total costs. Funds may be obligated at the outset of the program for later engineering, actual production, and deployment, but the larger and more complex the system, the more the actual spending will be concentrated in "out years"—and at that late stage it is possible to argue that we are already so far down the road with this system that it would be a pity to abandon it.

The consequences of the infusion of new funds in this manner may be, paradoxically, to reduce the overall effectiveness of the military establishment. More elaborate weapons may produce a more muscle-bound military, particularly if it is strapped for funds to practice how to use and maintain those weapons. On the nuclear side, greater accuracy and more destructive capacity, as in the MX missile (for which no survivable launching pads have yet been identified), may suggest to the Soviets that the United States is contemplating a first strike that could destroy a large proportion of their nuclear forces, heavily concentrated in fixed land-based silos. This fear may lead the Soviets to adopt a launch-on-warning policy which could all too easily trigger a nuclear exchange that neither side intended.

New nonnuclear weapon systems have greater firepower, speed, and complexity, but they may stretch the capacity of their operators and maintenance crews almost to the breaking point. The principal criticism of the new M-1 tank seems to be its frequency-of-repair record. The advanced electronic gadgets in the F-18 have been criticized as being less reliable than simpler but more rugged systems in earlier designs. And the recent naval battles off the Falklands have raised serious questions about the vulnerability of even the biggest and best-defended capital ships.

What remain unclear in this scenario are the national security and foreign policy objectives to be served by the additional expenditures. Clearly, this third scenario differs from the first in its continuing commitments to allies in Western Europe and the Far East, manifested through the continued presence of United States forces, as well as in other ways. It differs also in the commitment, somewhat imprecisely defined, to resist the takeover of Persian Gulf oil resources by an outside force, presumably the Soviet Union. And it differs in its determination to keep sea-lanes open for United States commerce. It is less clear how this scenario differs from the second scenario. Presumably, it would pursue the same objectives with more dollars to spend on them. But, for the reasons given above, there are grounds to believe that it might be no more effective.

We now come to the fourth scenario, sometimes characterized as the "military reform" or "not bigger or smaller, but better" scenario. It assumes some increases in the defense budget, but it focuses on selectivity, innovation, and reexamination of basic concepts and doctrines. As Senator Gary Hart, a leader in the Senate Armed Services Committee, put it recently: "If the Reagan Administration is serious about efforts to strengthen the military, it will have to look beyond the size of the budget."

The military reformers, who include a number of younger officers, members of Congress and congressional staff, and academic observers of the military scene, argue that the United States and its NATO allies can no longer afford to rely on military doctrine that emphasizes massive firepower rather than maneuver any more than we could continue to rely on nuclear superiority. The reformers question the value of giant aircraft carriers and associated carrier task forces over larger numbers of smaller, less expensive carriers, with vertical or short-takeoff-and-landing aircraft. They put more faith in versatility and reliability, rather than in high performance, as desirable airplane characteristics. They give high marks for technological ingenuity, but they put their greatest emphasis on innovations in military theory—and on the kinds of military education and career management that offer scope for the military theorist.

The reformers are detemined to change the situation that has led Paul Bracken and Martin Shubik to ask the question: "Has the Department of Defense become incapable of being intellectually surprised?"

Military reformers seem equally concerned to develop new techniques to deal with low-level violence, including international terrorism, as with conventional war. They speculate that the bureaucratic tendencies that seem to go with very large size in the military establishment may be the worst enemies of effectiveness, and they look for ways to offset those tendencies without sacrificing critical mass. It may be that, when the administration's military budget projections begin to meet increasing resistance on Capitol Hill and throughout the country, some of the reformers' proposals may find increasing favor.

The reform movement has not yet matured to the point where it divides in order to go off in several different directions. But "reform" cannot persist indefinitely as a unifying concept; changes can be made in different and inconsistent ways.

At this stage, at least three separate axes of development can be discerned:

The first axis is one of technological simplification; multipurpose weapons, ruggedly designed, relatively simple to operate and relatively (everything *is* relative) inexpensive per copy, are substituted for complex, expensive, and highly specialized systems, some of them said to be beyond the competence of today's recruits to handle. Also, since simpler weapons are less expensive, one can buy more of them, as with the smaller aircraft carriers advocated by Senator Hart.

Charles Hitch, the man who introduced program budgeting in the Pentagon under Secretary of Defense McNamara in the Kennedy era, likes to tell the story of the bronze spears and the iron spears. When iron spears were introduced, some thousand years before the Christian era, bronze spears were clearly superior; but bronze ones were so much more expensive that only one soldier in ten could be equipped with them, and the ranks of soldiers armed with the cheap iron spears won the battles and the war.

In the same vein, Eugene Fubini, former chairman of the Defense Science Board and a philosopher among defense engineers, observes that one shouldn't buy the "best" system, because it won't work under adverse conditions; one shouldn't buy the second best, because it will cost too much for what it will do; one should buy the third best.

The second axis is oriented toward sophisticated simplicity in weapons, like the Exocet missile, the single "smart bomb" that homed in on and sank the British destroyer Sheffield, 30 miles away in the South Atlantic. It has even been suggested that some of these weapons might be able to take the place of battlefield nuclear weapons, thus raising the nuclear threshold, above which there may be no stopping place until Armageddon is reached.

Such sophisticated weapons, it is argued, cannot be designed by a committee, but only by an individual or a small team free of institutional restraints. Such a team developed the Sidewinder heat-seeking missile which is so effective in aerial combat. The Sidewinder was designed and built by a small group of scientists in the defense laboratory at China Lake, in the Californian desert, working on their own time and with scrounged materials. A similar environment was to be found in the "skunk works," a completely separate division of the Lockheed Company, where Kelly Johnson designed and built the U-2 and the SR-71, the revolutionary reconnaissance aircraft.

The optimal conditions for developing weapons of elegant, sophisticated simplicity may not differ from the optimal conditions for producing weapons of rugged simplicity. The kinds of warfare they envisage, however, may be quite different. Sophisticated weapons may obviate the need for overwhelming firepower, and they may be launched and operated from well behind the front lines.

On the other hand, they put very heavy demands on what military shorthand labels C^3I: command control, communications, and intelligence. High level direction and coordination of sophisticated weapons, based on adequate and accurate intelligence, are essential to their effective use. It is a continuing complaint that the C³I function tends to be neglected in favor of more glamorous military concerns. Current anxiety about the adequacy of the function in a crisis, particularly where the use of nuclear weapons is threatened, has attracted new attention and new resources. But most reformers believe that even this increased attention is still insufficient.

In their most highly developed form, these weapons may introduce the concept of the automated battlefield in which technicians working in clean, well-lighted rooms engage each other in a kind of giant television war game, locating each others' remotely operated weapons by using remote sensors. The automated battlefield may never become a reality and, in any event, the concept does not cover the situation in which one side is unwilling to accept defeat according to the rules of the game. But long before one reaches the automated battlefield, the first two axes appear to diverge.

A third axis of development points toward a style of warfare that emphasizes maneuver over firepower.

"Maneuver warfare" has been defined, in legislation introduced by Senator Hart calling for a special study of the subject, as "envisioning an arms conflict as time-competitive observation-orientation-decision-action cycles in which the object is to destroy the enemy's cohesion by maintaining a consistently faster cycle." Reaching beyond the stilted language of the legislative draftsman, Senator Hart cited recent examples of maneuver war: the German blitzkrieg, General Patton's campaigns, the landing at Inchon in Korea, and "almost all of Israeli campaigns." "Maneuver," he explained, "is not simply a matter of moving, or even of moving rapidly. Maneuver means moving and acting consistently more rapidly than the opponent." The object is to be able to react more quickly, to go from observation to action faster than the enemy can move at each encounter, until he feels he has lost control of the situation. Strategy, doctrine, weaponry, and training are all focused on this ability.

Maneuver necessarily puts heavy emphasis on individual training, and consequently encourages experimentation with manpower
policy. One option is to place more emphasis on highly trained
cadre in the active duty forces and on experienced, regularly exercised reserves to fill out the units. With smaller operational units,
reserves should be more readily mobilized, and with the accent on
movement rather than firepower, experienced reserves might be
more valuable than more recently trained recruits. A recent report
by a group of National Guard generals that advocated, among other
things, the maneuver strategy, even proposed that National Guard
units choose and procure their own simple, inexpensive equipment.
Here the first and third axes intersect.

Maneuver warfare has historically had a particular attraction for
the side that is outnumbered or outgunned. It appeals to the David
in every citizen of the Western democracies, to the tradition of
Merrill's Marauders rather than the charge of the Light Brigade.
But one does not have to accept the proposition that the Soviets are
stronger in conventional forces to support the maneuver concept.

There is controversy over the application of the doctrine in
Western Europe and particularly in the north German plain. Since
NATO was put together, West Germans have insisted on a "forward
defense" so that their territory would not be overrun; a forward
defense is still official NATO strategy. But among those who are
exploring alternatives, the proponents of maneuver warfare are concerned that military planners are trying out some of their ideas but
with concentrations of firepower so great as to defeat the basic
purpose—and to scare off the Germans who don't want their entire
country to become a battlefield.

The fourth axis of reform is organization—or rather reorganization. Some of the more organization-minded of the reformers set a
good deal of store by a greater degree of unification among the four
military services. They point out that the compartmentation of warfare into land, sea, and air wars, with a special amphibious role for
the marines, is quite anachronistic, and they observe also that such
integration as exists in the operational commands and the Joint Staff
is more formal than functional. So long as the members of the Joint
Chiefs of Staff look to their service staffs rather than to the Joint Staff

to prepare their positions, and so long as officers serving in joint commands look to their respective services for promotion, unification is a very pale reality. No one has seriously proposed a single service since the Symington Report a decade after World War II. What the military referred to as "purple suits" are not likely to replace army khaki and navy blues for the foreseeable future. But General David Jones, who retired as chairman of the Joint Chiefs of Staff in the summer of 1982, publicly advocated, as he was leaving office, organizational changes to make the Joint Staff truly joint, and to strengthen the hand of the chairman. General Edward Meyer, chief of staff of the army, has gone even further; he advocates divorcing the members of the Joint Chiefs from their roles as heads of their respective services so that they form a sort of military advisory council to the secretary of defense and the president. Both Jones and Meyer would strengthen the hand of the unified commanders in the field, breaking down interservice barriers to permit tighter coordination (a special requirement for maneuver warfare). There are no indications, however, that the services are prepared to surrender any of their traditional prerogatives. It seems likely that organizational changes will still be debated long after major changes in strategy and force structure have been effected.

The reformers are evidently not all moving in the same direction. But at least they are talking to each other, so that diverging paths can come back together. And as suggested, a rising defense budget creates a better climate for reform, as was last demonstrated in the early 1960s.

No account of alternative futures would be complete without at least mentioning some of the more far-out alternatives that appear from time to time in newspaper feature stories. The simple fact is that (like nuclear wars) space wars, biological weapons wars, or even chemical weapons wars are not a reasonable method by which one government can work its will on another.* Perhaps the strongest conclusion that one can reach about the future of military establish-

---

*This is not to deny the vital importance of orbiting satellites in verification, particularly of nuclear weapons deployment, nor the value of a minimal chemical warfare capability so that neither side can force the other into the inconvenience of protective measures without being compelled to the same measures itself.

ments in general is that, almost in the proportion that they become more frightening, they become less effective instruments of national policy.

At the other end of the spectrum of violence, the military may become more involved in learning to cope with low-level internal violence and specifically with the phenomenon of terrorism, from which the United States has thus far been almost completely spared. Federal troops have been called on only rarely to intervene in domestic disorders: in only eighteen major incidents over the 190 years since that use was authorized by Congress has intervention been necessary. But a series of terrorist incidents, or even one major disruption of daily life in one large metropolitan area—an accomplishment well within the reach of a professional terrorist group—could make significant demands on the military to help provide essential supplies and services and to assist the state and local police in maintaining order.

The scenarios for the future of the military establishment cover an even broader spectrum than the present expectations of the American people for the military. To the extent that the military budget remains at or above present levels, in a period of fixed real incomes and declining psychic incomes for most Americans, the pressure on the military to operate more efficiently and effectively, and on the civilian leadership to define and refine national security policy, will be particularly severe. It remains to be seen whether these pressures will be great enough to overcome the centrifugal tendencies of what is almost certain to continue as the largest bureaucracy in the country, and to force the country to decide what it wants of its military.

VIII

Arms Control in the Eighties

IN ASSESSING ALTERNATIVE futures for the military establishment—
and indeed for the human race—perhaps the most critical factor is
the potential for effective arms control. Each superpower feels the
need to maintain a military force adequate to deter the other side
from interfering with its vital interests. This is not to suggest that
United States and Soviet military strategies are mirror images of
each other. In fact, there is a good deal of evidence to the contrary.
But there are powerful reasons why both governments seem to have
difficulty in exercising reasonable control over the size of their military establishments. Estimates of what it takes for effective deterrence are notoriously imprecise. Because of the extended timetable
for the weapons development process, major weapons development
decisions have to be made a decade before the shaping of the international political constellation in which they are expected to play a
role. Both military establishments have to be so large, in order to
meet minimum security requirements, that they tend to outweigh
other elements in the formation of national policy.

In the United States, our national policy on the size of the
military establishment tends to veer between extremes of concern
over military risks on the one hand and over budgetary costs on the
other. These cyclical fluctuations are imposed on a secular trend
toward increased spending. In the Soviet Union, the indications we
have are that the budgetary trend for its military establishment is

still inexorably upward.* In the absence of effective arms control
agreements, those trends are likely to continue with consequent
increases in cost and instability and in the ultimate danger of nuclear war. After spending more billions of dollars and billions of
rubles, both sides are likely to be less rather than more secure.

Arms control agreements can and do serve other purposes as
well. In a world where arms competition is clearly not limited to the
superpowers, agreements can restrain competition in the size and
shape of other countries' armed forces. They can inhibit the spread
of the most dangerous and destructive weapons beyond the present
members of the nuclear club. They can rule out certain kinds of
warfare (e.g., chemical, biological), and they can even attempt to
prohibit possession of the means to conduct these kinds of warfare.
They can make hostilities by accident or preemption less likely—or
they can simply facilitate crisis communications, as in the case of the
hot line agreement between the United States and the Soviet Union—the first American-Soviet agreement in this area—entered into
after the Cuban missile crisis.

Of course, agreements themselves are not automatically self-
enforcing. They must be *verifiable*, in that each party must be able
to discover promptly if the other side is in violation of the agreement. They must be *equitable*, in that each side must believe that
the agreement is fair, or sooner or later a way out of the agreement
will be sought. And, they must be *practical*, in that any party
tempted to abandon the agreement must be convinced that any
advantage gained thereby would be short-lived and not decisive.

Arms control need not depend entirely on formal agreements.
One power may take temporary unilateral action, as President Ford
did in 1975 by directing the dismantling of America's lone authorized antiballistic missile site, and may wait just long enough so
as not to put itself at a significant military disadvantage to see

*There is at least moderate agreement in the defense research community that
Soviet military procurement has grown at the fairly steady rate of 4 percent a year (in
constant dollars) over the past ten years, and that Soviet defense spending absorbs a
fairly constant 11–13 percent of their gross national product, while the United States
percentage has varied between 5 and 10 percent of a much larger GNP.

whether the other power will follow suit. The calculation is a tricky one, but the rewards for all the parties can be considerable. Or one power may choose a particular weapon system or force structure that it believes will protect its vital national interests in a fashion less likely to provoke a counter development in weapons or a situation of greater instability. Again, the calculation is a tricky one.

A major piece of President Carter's arms control program was the attempt to cut back on United States sales of arms to other countries, primarily Third World countries. Although the program was extremely modest in concept, it was essentially abandoned by the end of the Carter administration, and the 1980 foreign aid bill even relaxed government controls on arms sales by private companies. It only remained for the Reagan administration to announce an explicit reversal of the Carter policies, including the "leper" directive discussed in chapter 3 above.

The early history of arms control attempts involving the United States is short and not particularly sweet. The Washington (1921–22) and London (1930) Naval Arms Limitation Treaties between the two World Wars served only temporarily to codify the status quo. The Kellogg-Briand Pact (1928), which renounced war as an instrument of national policy, will go down in history only as an extraordinary case of rhetorical self-delusion. It demonstrated clearly the futility of proscribing behavior without imposing obligations on the participants or providing means of enforcement. The use (but not the manufacture) of chemical and biological weapons was banned by a 1925 Geneva agreement—although not ratified by the United States until fifty years later. The manufacture of biological weapons was banned by a multilateral convention signed by 113 countries between 1972 and 1976. A pall was cast over the agreement by an outbreak of anthrax in the Russian city of Sverdlovsk in 1977, arguably caused by an accident in a biological weapons plant, and then by strong indications of Soviet use of such weapons in Southeast Asia.

Even with the advent of the nuclear weapons age, serious arms control negotiations were slow to get under way. The Baruch plan for international control of all atomic energy was never accepted by

the Soviets and national development of nuclear weapons pro-
ceeded apace in the United States, the Soviet Union, and the
United Kingdom, and then in France, and China. The first efforts to
achieve some measure of international agreement on the control of
nuclear weapons (apart from the 1963 United States-Soviet Union
hot line agreement) focused on a proposed ban on nuclear weapons
tests, without which development of new weapons was thought to
be extremely difficult. The test ban proposal received particular
impetus from popular concern about the fallout from tests in the
atmosphere producing serious health and genetic consequences. An
American-Soviet moratorium in 1958 was terminated with resumption of Soviet testing in 1961, and the United States followed suit
shortly thereafter by resuming its own testing.

A major accomplishment of the Kennedy administration was
the 1963 Limited Test Ban Treaty, which succeeded in banning
nuclear tests in the atmosphere, underwater, in outer space, or in
any other environment if the explosion would cause radioactive
debris to be present outside the borders of the country conducting
the explosion. Neither France nor China became a signatory to this
treaty. The ban failed to include underground tests since agreement
on that issue had broken down on the specifics of the on-site inspec-
tions considered necessary for verification purposes. Indeed, as a
condition of obtaining the endorsement of the Joint Chiefs of Staff,
the administration was forced to agree to an intensified program of
underground testing that has since been able to accomplish virtually
all the technical military objectives that could have been accomplished by testing in the atmosphere. Further, the test ban put no
limitations on the testing of nuclear delivery vehicles, so that developments like the MIRV and the cruise missile were not affected.
Two supporting agreements—the 1974 Threshold Test Ban Treaty
and the 1976 Treaty on Underground Explosions for Peaceful Purposes—reflect a willingness on the part of both the Soviets and
ourselves to achieve the marginal benefits of limiting the conduct of
nuclear detonations without making the total commitment necessary for a complete test ban. Neither of these agreements has yet
been ratified by the United States Senate.

The second major treaty limiting nuclear weapons was the 1968

Non-Proliferation Treaty (NPT), in which 58 nonnuclear states pledged themselves not to develop or acquire nuclear weapons. In addition, three of the five nuclear powers—the United States, the Soviet Union, and the United Kingdom—pledged themselves not to transfer nuclear weapons to others and to exert their best efforts to achieve mutual reduction of their own nuclear arsenals. The results of the treaty to date are somewhat in dispute. It has not discouraged India from acquiring a nuclear weapons capability, and there is some reason to believe that Israel and even South Africa may also have done so, although none of these parties has declared, or even admitted to acquiring, such a capability. The second five-year review conference for the treaty, in 1980, broke up without achieving agreement on any major point. But the fact remains that few would have believed ten years earlier that, within the decade, there would be no new declared members in the nuclear club, and only two or three possible clandestine members.

The future of the Non-Proliferation Treaty, and the prospects for avoiding further extensive spread of nuclear weapons to other powers (and eventually to political factions and terrorist groups), depends significantly on the ability of the nuclear superpowers to achieve some degree of mutual restraint in their own nuclear arsenals. Related agreements to ban nuclear weapons and weapons of mass destruction from the seabed (1971) and from outer space (1967) generally have been regarded by the 110 non-nuclear NPT signatories as less than major efforts.

The concerns of these nonnuclear states have focused principally on the Strategic Arms Limitation Talks (SALT) between the United States and the Soviet Union. The talks were first proposed at the 1967 summit meeting between President Johnson and Soviet Premier Kosygin in Glassboro, New Jersey. After a number of delays, the initiation of the SALT process eventually resulted in the SALT I agreement in 1972 and the SALT II agreement, signed in 1979 but withdrawn from Senate consideration and still not ratified by the United States Senate at this writing. Despite United States failure to ratify, SALT II is mutually observed, as explained in more detail below.

SALT I consisted of two parts: (1) a treaty banning the development of antiballistic missiles (except for two small sites on each side), and (2) an interim agreement limiting the numbers of offensive missile launchers—strategic bombers, silos for land-based missiles, and launching tubes on ballistic missile submarines.

The ABM treaty—which enjoins both the United States and the Soviet Union "not to employ ABM systems for the defense of its country and not to provide a base for such a defense"—eliminated a weapon that might have had a seriously destabilizing effect on the nuclear balance. Perhaps more to the point, the treaty made unnecessary expensive efforts to develop a weapon in which neither side had much confidence but without which had feared being left behind in a technology race.

The interim agreement on strategic offensive weapons paved the way for negotiation of a permanent cap on strategic nuclear weapons of all kinds, although it passed up the opportunity to curb the development of perhaps the most dangerous destabilizing missile: the land-based MIRV'd ICBM. Achievement of this permanent cap was first sought at Vladivostok in 1974, unsuccessfully, and a final agreement was not completed until five years later after bipartisan efforts by three United States administrations.

The SALT II agreement, even in its unratified state, represents a major milestone in nuclear arms control negotiations for a number of reasons. It takes as the unit of measurement nuclear weapon launchers rather than nuclear warheads, since the number of launchers can be observed and verified by so-called national technical means (NTM)—including the extensive use of photographic satellites—without more than passive cooperation from the other side. Thus each side will know if the other side is cheating. The agreement places an overall ceiling on strategic weapon launchers, limiting both parties to the same total number of launchers (2,250), while permitting each side to mix various types of launchers—bombers, land- and sea-based missiles—in accordance with its own strategic needs and doctrine. The agreement contains specific commitments not to interfere with the NTM employed by the other side. It limits the modernization of strategic nuclear forces by permitting each side only one new ICBM, and by forbidding the addition of more warheads to a MIRV'd missile, or even the production of a mobile strategic missile. All these limits are fixed in ways that are adequately verifiable. The agreement deliberately postpones the troublesome questions of controls over ground- and sea-based cruise missiles.

The treaty covers only strategic weapons—weapons that can reach the territory of one of the parties from the territory of the other (or from submarines presumably in international waters). One of the controversial issues is the exclusion of the Soviet Backfire bomber, which, while clearly intended for other purposes, could theoretically be deployed on one-way missions against continental United States targets. Negotiations designed to control "theater" nuclear weapons—the Intermediate Nuclear Force Talks—were initiated in Geneva, Switzerland, on November 30, 1981.

These negotiations exclude the so-called battlefield nuclear weapons, at the lower end of the scale, and intercontinental, or strategic weapons, at the upper end of the scale. Negotiations were provoked by the Western reaction to continuing Soviet deployment of the SS-20, a MIRV'd, mobile missile with improved accuracy that is being added to earlier generation SS-4s and SS-5s and in much larger numbers. The initial European reaction to the deployment of the SS-20 and the Soviet Backfire bomber was to call either for their limitation by including them within the scope of the SALT negotiations or for the introduction of additional United States theater nuclear weapons—specifically, ground-launched cruise missiles and extended-range Pershing II missiles. These would supplement the Europe-based F-111 bombers, and the nuclear-armed submarines and carrier-based aircraft, some of which can strike Soviet targets.

But the NATO decision to deploy these new weapons was made over the vigorous objection of politically important elements throughout Western Europe. It was accepted only on the understanding that the United States would pursue a parallel effort to negotiate with the Soviets a substantially lower level of theater nuclear weapons (other than battlefield nuclears) that would make the additional NATO deployments unnecessary. The Soviets attempted to exploit the European reactions by proposing a freeze on both sides. While such a freeze would clearly preserve an asymmetric situation (if one looks only at Europe-based weapons), it had considerable propaganda appeal in Europe. The situation was not improved by the United States announcement of its decision to go ahead with production of the so-called neutron bomb.*

On November 18, 1981, President Reagan countered the Brezhnev proposal with a more drastic, if equally nonnegotiable American proposal. This "zero option" called for:

—Dismantling of the 600 Soviet intermediate-range ballistic missiles in exchange for United States cancellation of a plan to deploy 572 Pershing II and cruise missiles in Western Europe.
—Substantial reduction of Soviet and American strategic nuclear weapons to equal and verifiable levels.
—Equalization of conventional forces in Europe at lower levels.
—Development of measures that would lower the risks of surprise attack and war by miscalculation.

Implicit in the Reagan proposal (whether intentional or inadvertent) was a recognition that meaningful resolution of the theater nuclear question is possible only in the context of a resumption of the SALT process, in some form. Without a cap on strategic systems, the Soviets could increase the number of their "strategic" launchers and train some of them on Europe, as they have in the past. The lack of a strategic nuclear arms agreement would create a climate of uncertainty and instability that would make any other nuclear arms agreement even more difficult to achieve and maintain.

After candidate Reagan campaigned against ratification of SALT II, the Reagan administration officially disowned the treaty as "fatally flawed"—although it did not make altogether clear the nature and significance of the claimed flaws. Nevertheless, both sides

---

*The bomb, described in chapter 2 above, was predictably attacked by some Europeans as a "capitalist" weapon that would destroy people without injuring property. The deeper concern was that its capability to knock out enemy forces advancing into NATO territory without as much collateral damage to that territory might reduce the size of the "firebreak" between the employment of conventional and of nuclear weapons in a crisis.

continued to observe the provisions of SALT II, with the result that only the Soviets appeared to benefit from American failure to ratify: they would have been required to dismantle several hundred of their weapons on ratification, and the United States would not have been required to dismantle any.

In place of SALT, the new administration proposed to initiate Strategic Arms Reduction Talks, optimistically dubbed START, that would emphasize significant cuts in strategic nuclear arsenals rather than a cap on present deployments. During President Reagan's first year in office, Soviet involvement in the suppression of Poland's struggle for democratic reforms, continued fighting in Afghanistan, evidence in Southeast Asia of Soviet violation of international agreements against the use of chemical and biological weapons, and attempts on the United States side to lay every instance of violence in the world at the Soviet doorstep did not create a propitious climate for START to begin.

Meanwhile, away from the negotiating tables in Geneva, Vienna, Helsinki, and Madrid, and outside the windowless conference rooms in the Pentagon and the State Department, a great debate was building in the country on what can and should be done to control the nuclear genii.

The last great nuclear debate took place in the 1960s.* It began with an election year argument about whether there was or would be a "missile gap" between the United States and the Soviet Union, an argument that was the mirror image of election year arguments in 1980.

The early 1960s were the era of children learning to hide under their school desks in civil defense drills, mothers worrying about strontium 90 in the milk supply because of the fallout from atmospheric testing, tension with the Soviets over the status of Berlin, and then the Cuban missile crisis, when the prospect of nuclear war

---

*An even earlier debate, immediately after the first (and only) use of nuclear weapons in war, by the United States at Hiroshima and Nagasaki, addressed the question of which governmental mechanisms, national and international, should be used to control these weapons. With an assist from the Soviets, those who favored national over international control won out.

seemed almost close enough to touch. The debate that paralleled these concerns culminated in the negotiation and ratification of the 1963 Limited Test Ban Treaty, described above. After that, the vivid and frightening symbolism of the mushroom cloud disappeared from the world scene, and public concern died down. It flared up again in the late 1960s when the Johnson administration proposed to deploy antiballistic missiles, bringing the idea of nuclear war into everybody's backyard. That debate was eventually resolved in the ABM Treaty, and meanwhile popular attention was diverted to the unconventional use of "conventional" military power in the tragic but nonnuclear war in Vietnam.

The first proposals by the Carter administration for the deployment of the new MX missile, more accurate and more powerful than the Minuteman, marked the beginning of a new great debate on nuclear weapons. Somehow the SALT negotiations had not provoked any extensive controversy outside the corridors of power and the halls of Congress. The decision to deploy the new missile in a network of interconnected silos spread across several western states, with missiles being moved back and forth around a kind of giant racetrack, or along a special-purpose railroad or tunnel, aroused unexpected and vociferous opposition from conservative ranchers and local business people. In a public forum in Utah, attended by the undersecretary of the air force and broadcast live on national television, a man who looked like a ranch owner out of central casting got up to observe that he had heard about the racetrack mode and the railroad mode and he thought that some consideration should be given to the commode for the MX. The assembly dissolved in laughter.

The concerns of the cattlemen and sheepherders mingled with the concerns of the conservationists, some of whom were already stirred up by controversies over the safety of nuclear power plants and the disposal of nuclear wastes that would remain poisonous for thousands if not tens of thousands of years. It happened also that a portion of the territory where it was proposed to deploy the MX was part of what geologists call the "overthrust belt," where the prospective exploitation of shale oil and coal deposits was already believed by many to threaten both the natural and the man-made environment of the West.

These concerns abated somewhat during the hiatus after the Carter administration abandoned the effort to have SALT II ratified, and television focused instead on the Soviet invasion of Afghanistan and the detention of American hostages in Iran. The extensive campaign toward the end of the Carter administration to develop public interest in the SALT process could not seem to attract public attention. But beginning about midway through the first year of the Reagan administration, popular attitudes seemed to shift in almost glacial fashion. A group of physicians, concerned about the prospect of nuclear war as a medical problem of unprecedented and unmanageable proportions, organized Physicians for Social Responsibility. The organization aims essentially to educate Americans about what a nuclear exchange would mean in medical terms, in trauma, burns, radiation sickness, and starvation, all in the face of complete disruption of medical services. A former National Security Council staff member, Roger Molander, gave up his career as a national security analyst to found an organization called Ground Zero, dedicated to explaining, in graphic terms, the devastation that would result from a nuclear exchange between the United States and the Soviet Union. The movement grew and spread. Local Ground Zero organizations sprang up all over the country. A *New Yorker* magazine staff writer, Jonathan Schell, published a series of articles on the human consequences of a nuclear exchange. He rejected the phrase "nuclear war" since he maintained that a nuclear "war" would be unlike any other war that has ever been waged, and could not accomplish any of the traditional political objectives of war. The articles, later published in book form, were an instant success, rather like novelist John Hersey's article on Hiroshima, which had taken up an entire issue of the *New Yorker* some 35 years earlier.

Some of this mounting concern was prompted by several careless or inadvertent remarks of President Reagan, suggesting the possibility of a limited nuclear war in Europe and indicating that he was not altogether familiar with NATO doctrine on the delicate question of when nuclear weapons might be employed. The situation was further exacerbated by conflicting statements of Secretary of State Haig and Secretary of Defense Weinberger on United States nuclear strategy; the conflict was only papered over by the White House.

A good deal of the grass roots debate and organization focused on the idea of a general nuclear freeze (not to be confused with the Brezhnev European freeze proposal), calling for a mutual American-Soviet cap on the production, deployment, and modernization of the existing arsenals of nuclear weapons. Popular referenda on the freeze idea were added to the ballot across the country, from New England town meetings to a statewide initiative in California, which, at this writing, seems likely to be carried in November 1982 by a significant margin. A more sophisticated version of the freeze proposal, calling for the immediate initiation of the American-Soviet negotiations "to decide when and how to achieve a mutual and verifiable freeze . . ." was introduced in the United States Senate by Senators Edward Kennedy and Mark Hatfield. The resolution was cosponsored by 22 other senators, and 166 members of the House of Representatives, as well as by a long list of national security experts, religious leaders, and private citizens.*

At about the same time that the Kennedy-Hatfield resolution was introduced in the Senate, four eminent citizens, McGeorge Bundy, President Kennedy's special assistant for national security affairs, George Kennan, former ambassador to the Soviet Union, Robert McNamara, secretary of defense under Presidents Kennedy and Johnson, and Gerard Smith, chief negotiator of the SALT I treaty, published an article in the erudite and even sometimes stuffy *Foreign Affairs*, the quarterly journal of the Council on Foreign Relations, suggesting that the United States should adopt an explicit policy of no first use of nuclear weapons. Such a policy would constitute an abandonment of the official NATO doctrine that Western Europe was somehow protected, even if its nonnuclear forces failed to stem a (nonnuclear) Soviet invasion, by the American nuclear umbrella. The article added fuel to the nuclear debate, even more

---

*By way of full disclosure, the senior author must point out that he was among the cosponsors.

so than Ambassador Kennan's proposal, a year earlier, for a United States-Soviet agreement on a 50 percent cut in nuclear forces. Opponents of the no-first-use idea argued that the possibility of first use was necessary, even as a bluff, to overcome Soviet conventional superiority in Europe. The proponents replied that without a no-first-use doctrine NATO would never pay proper attention to its nonnuclear forces, and that in any event the nonnuclear imbalance was neither so clear nor so significant as the other side claimed it was.

At the end of May 1982, the Palme Commission, an international commission on disarmament and security issues chaired by Olof Palme, former Swedish prime minister, and including Cyrus Vance, secretary of state in the Carter administration, and Georgi A. Arbatov, the Soviets' principal expert on the United States, issued its report. Its most newsworthy recommendation was for both sides to remove all tactical (or battlefield) nuclear weapons from Central Europe by 1983—again to widen the firebreak between conventional and nuclear weapons.

Other nuclear arms control ideas surfaced, or rather resurfaced in the general debate: giving priority to completing the negotiations on a comprehensive test ban treaty, initiating negotiations on a cutoff in the production of weapons-grade nuclear material, and a focus on mutual elimination of the most unstable weapons, namely the MIRV'd land-based missiles, which are most likely to be employed in a preemptive attack because they are most vulnerable themselves to a first strike ("Use 'em or lose 'em" was the phrase used to describe their situation).

Still another proposal, by Jeremy Stone, director of the Federation of American Scientists (appropriately from a mathematician turned political organizer), called for an annual percentage reduction, mutually agreed, in all nuclear weapons. The proposal suggested a gradual decrease in the rate of reduction so that enough weapons remain on each side to avoid the situation in which the side with a small advantage could upset the strategic balance. A simpler variant of the same approach came from a distinguished retired admiral, who suggested that the United States and the Soviet Union agree that each side turn in a specified number of weapons, of its own choice, starting, say, with an experimental 50 on each side, and increasing in numbers as both sides gained confidence in the process. He pointed out that, as an additional and important benefit of the proposal, each side would be likely to turn in its most unstable weapons first. Several veteran arms control experts even suggested that the time might be ripe to revive the ratification debate on SALT II, and a joint resolution to that effect was introduced in Congress, but failed of passage by a narrow margin.

The consensus among those professionally and politically concerned about arms control was that with all the technical difficulties attendant upon negotiating a complete freeze as the next step in the arms control process, rather than negotiating deep cuts in selected weaponry, the freeze movement still provided the essential popular impetus that might bring technical negotiations to a head. The freeze movement, the Ground Zero Organization, the Committee for National Security, the professional groups like the organized physicians, and later (ironically) the organized lawyers, represented a kind of life-force, the word made flesh.*

It was also argued that the freeze proposals call attention to the fact that "modernization" of weapons, by increasing accuracy and destructive power, tends to greater instability because each side's weapons may become more vulnerable, as suggested above in the case of MIRV'd ICBMs.

As of this writing, President Reagan has responded to these concerns in a major speech delivered to the graduating class of his alma mater, Eureka College, on May 9, 1982, in which he called for a two-stage negotiation with the Soviets. The first stage would be focused on what he described as "the most de-stabilizing systems— [land-based] ballistic missiles." He proposed to use warheads as well as launchers as the units of account and to try to reduce to "equal ceilings at least a third below current levels," of which no more than half would be land-based.

---

*The senior author recalls his last conversation with President Kennedy, a few days before the assassination, when Kennedy spoke of the untapped reservoir of popular concern, which he had found in his last trip around the country, on the issue of war and peace.

Since the United States has an edge over the Soviet Union in numbers of warheads, according to most people's arithmetic, the proposal should have some appeal to the Soviets. On the other hand, some three-quarters of the Soviet strategic nuclear capability is in land-based missiles, and therefore the Reagan proposal would require major reduction or shifts in their missile basing. Further, counting warheads is not like counting launchers, which can be observed from satellite photography. SALT II contains counting rules for MIRV'd (multiple warhead) missiles, based on flight testing, which can be observed. In brief, missiles are assumed to have the maximum number of warheads for which they have been flight tested. The alternative would be a quite intrusive form of on-site inspection, which the Soviets have traditionally resisted, although there has been some recent indication of relaxation in this area.

In his speech, President Reagan proposed as the second phase "an equal ceiling on other elements of our strategic nuclear forces, including limits on ballistic missile throw-weight." This negotiating sequence apparently represents a victory for those within the administration who favored an approach more likely to produce positive results, influenced perhaps by the popular pressure generated by the freeze movement. Both in his speech, and immediately afterwards, President Reagan continued to reject the idea of ratifying SALT II, although his administration continued tacitly to observe its limitations.

Then, in a Memorial Day 1982 speech, he pledged to observe the SALT II limitations so long as the Soviet Union did likewise. Meanwhile, in the same Memorial Day remarks, he announced agreement with the Soviets that the START negotiations would begin on June 29. It remains to be seen how the program he has proposed for negotiations will be executed. Initial calculations suggest that the specific cuts proposed by the Reagan administration would, if anything, increase the vulnerability of land-based missiles on both sides.

Meanwhile, the Reagan administration has announced that it is abandoning negotiations on a comprehensive test ban treaty, and it is "redefining" verification procedures for two signed but unratified Threshold Test Ban Agreements referred to above. There are no indications of any disposition on either side to resume negotiations barely begun in the Carter administration on an agreement to prohibit antisatellite weapons, and the United States apparently plans to deploy such weapons by 1987. Twenty-four countries have signed an agreement banning nuclear weapons from the Latin-American-Caribbean region (Argentina, Brazil, and Chile have not yet ratified). But real progress is unlikely until the SALT process has visibly advanced to a new stage.

If actual progress on nuclear arms control has been disappointing, progress on nonnuclear arms control has been even more so. The margins of the issue were gingerly touched on during the sixties with multilateral agreements to exclude any measures of a military nature from Antarctica. The centerpiece of nonnuclear arms control negotiations since 1973 has been the so-called Mutual and Balanced Force Reduction talks conducted in Vienna between the NATO allies and the Eastern bloc forces that face each other across the north German plain. These negotiations were originally conceived by NATO as an alternative to the efforts of then Senator Mike Mansfield, the Democratic majority leader, to force a unilateral withdrawal of United States troops in Europe. They have been hung up for a period of years on differences about the numbers of Eastern bloc troops in the area under negotiation, and there are no signs of an imminent breakthrough.*

Another set of United States-Soviet bilateral negotiations on naval forces in the Indian Ocean was broken off by the United States after the 1978 increase in Soviet military activity in the area. A third set of negotiations, on a United States-Soviet agreement to ban the production of chemical weapons (use of these weapons being already banned under the 1925 Geneva Convention) is effectively stalled by problems of verification. In the meantime, evidence mounts of Soviet use of chemical weapons in Afghanistan, and the United States Congress is on the verge of appropriating funds to

---

*During his June 1982 European trip, President Reagan did, however, move the talks at least a half-step ahead by finding a way to give the Soviets greater assurance that reductions on the NATO side would include West German as well as United States troops—a matter of particular concern to the Russians.

manufacture so-called binary weapons: projectiles containing two separately harmless substances which combine only when the projectile is fired, spreading deadly nerve gas on the target. A fourth set of bilateral talks on arms sales to Africa, Asia, and Latin America has shown no signs of progress after getting off to a very slow start in late 1977—and United States military sales have taken a new lease on life in the Reagan administration, as described in chapter 4 above.

A somewhat more promising negotiation has developed out of the Conference on Security and Cooperation in Europe. This conference, which first met in Helsinki in 1975, produced the Helsinki Final Act, adhered to by 35 nations, including the United States, the major European powers, and the Soviet bloc. While the portions of the Helsinki accords that deal with human rights have been notably ignored by the Soviets, continuing talks on the so-called Basket I issues of military security in Europe, including the prevention of surprise attacks and other "confidence-building measures" have produced a number of minor but useful agreements. The areas of accord cover mutual notification of maneuvers and troop movements, which should make a significant contribution not to reducing the size of military forces but to reducing the danger of confrontation escalating into armed conflict.*

Yet another forum was proposed by France in 1978, a conference on disarmament in Europe "from the Atlantic to the Urals." But despite some expressions of interest from the Warsaw Pact countries and West Germany, the conference has not yet been convened.

If all of this presents rather a mixed picture, there are at least three reasons why the picture may brighten over time:

First, arms control has become a respectable activity; until quite recently, it was barely so. Then secretary of defense, Harold

---

*Conference participants accepted a provision that specifies that notification be given at least 21 days in advance of any military maneuver involving more than 25,000 troops that takes place in Europe or, in the case of countries extending beyond Europe (the Soviet Union and Turkey), 250 kilometers inside their European borders. Participants also agreed voluntarily to exchange observers at these maneuvers, and to give notification of smaller scale movements.

Brown, in his 1979 military posture statement, observed that although "the trend in world armaments is upward, . . . security and stability can be better maintained by ceilings on, and reductions in, both nuclear and conventional capabilities," providing, he made clear, "they are specific, equitable and verifiable." As John Newhouse, a former assistant director of the Arms Control and Disarmament Agency, has pointed out, "the change worth noting is that arms control, instead of being occasionally serious, as generally in the past, appears now to have become permanently and unavoidably embedded in our national policy process."

Second, other nations are coming to grasp the importance and, indeed, the vital necessity of arms control. Although the Soviet Union still engages in a continuing series of propaganda exercises on the subject, it has also demonstrated its willingness to engage in serious negotiations looking toward concrete results. The French, who had remained aloof from the arms control and disarmament scene for a number of years, have come into the international arena with a whole series of proposals and have chosen to occupy their hitherto empty seat on the recently expanded multilateral Commission on Disarmament in Geneva. Even the Chinese have sent observers to the commission's meetings. Other governments as diverse as those of Britain, Yugoslavia, Rumania, India, and Pakistan have indicated active interest in the subject, focusing particularly on the United Nations Special Sessions on Disarmamant.

Third, the United States and the Soviet Union have begun to develop and perfect international machinery to do the dull but essential daily business of arms control verification, through institutions like the Standing Consultative Committee established under SALT I, which has, at least until recently, dealt so effectively with the delicate and difficult problems of treaty compliance.

These three factors are necessary but not sufficient conditions for effective arms control as a practical constraint on uncontrolled expansion of the United States and Soviet military establishments. The fact is that through the decade of the seventies the United States stock of alert nuclear weapons increased by 6,000 and the Soviet stock by 5,000. Three additional conditions are required:

In the first place, the objectives af arms control need to be separated from the objectives of general disarmament. This separation must be accomplished in order to avoid frustrating the doves and frightening the hawks. One can assume, as seems unhappily reasonable, that nation-states will continue for the foreseeable future to insist on maintaining the threat of military force in the international arena as an adjunct of national sovereignty. But it does not follow that they will refuse to agree on equitable, practicable, verifiable limitations on the instruments of military force in order to reduce the risk of actual conflict and the costs of deterring conflict.

Similarly, arms control must amount to more than ratification of preexisting decisions on weapons development and deployment. As one United States weapons analyst observed in the course of an academic seminar on arms control: "I can't think of a good weapon system we'd be willing to bargain away." But this observation begs the question, "What is a 'good' weapon system?" If a "good" system is one that is, among other virtues, consistent with reasonable arms control objectives, then the analyst's observation does not necessarily trivialize the arms control process. If he meant his observation more narrowly, and if he was correct, then arms control is still in trouble.

There is some reason to hope that the process is being internalized within the military establishment. Planners are speculating about the probable reactions of potential antagonists to proposed new United States systems. Proposals to deploy new systems are coupled with proposals, more or less serious depending on the circumstances, for negotiations to make those deployments unnecessary.

Lastly, the entire process depends on the leadership and the commitment of the president. Because arms control runs counter to the short-term parochial interests of the military bureaucracy—and those short-term parochial interests tend to override the individual concerns of even the most conscientious individuals in the system—it can only succeed when it has the vigorous support of the highest civilian leadership. Absent that support, negotiations may be resumed and even expanded, but positive results will not be forth-coming.*

Presidential commitment, in turn, will be a function largely of popular concern. The astonishing growth of the freeze movement and related developments in the United States have already been described. The European antinuclear movement has a breadth and depth that will not allow it to be dismissed as irresponsible just because its extreme fringe embraces unilateral disarmament. The masters of the Soviet Union may give little heed to public opinion, but at least the crisis in Poland, the continuing drain on their econ-omy caused by Afghanistan and Cuba, and the miserable state of the Soviet economy itself should suggest to them a pressing internal need to curb the appetite of the so-called metal eaters of the Soviet military-industrial bureaucracy.

Beyond the horizons of current arms control efforts lies the distant prospect of a world in which the rule of law does in fact prevail, and, even without general disarmament, the United States military can be converted from a war-fighting to a constabulary force.

The military constabulary function is still an unhappily vague concept, which has been much discussed but little defined. Were the Roman legions constabulary? Or the nineteenth-century British army operating on the borders of India? Or General Custer's forces at Little Big Horn? Or, for that matter, the United States forces at Wounded Knee? In a world of separate sovereignties, where two sovereignties clash, which side is the policeman?

For our present purposes, we can define a constabulary force as one that considers war both as an interruption of its normal duties and as an indication of failure in a system of relations in which it has a significant continuing responsibility. Both elements are important because a primarily war-oriented force could still realize that wars

*It is a paradox of politics that President Carter, who was strongly supportive of arms control, chose not to appear at the First Special Session on Disarmament of the United Nations General Assembly, in 1978, apparently for fear it would damage the chances of SALT II ratification; yet President Reagan, who has been a reluctant convert to arms control, has, at this writing, announced his intention to appear at the Second Special Session in June of 1982.

are exceptions to the normal course, or that it had some peacekeeping responsibilities. There is no black and white distinction between the two functions, particularly at a time when the United States has not been involved in a declared war for more than thirty years, and the second biggest war of the period was explicitly labeled a "police action."

In this situation, military attitudes are as important as military functions. Soldiers, by definition, are armed men. Yet thoughtful soldiers cannot help wondering about their future role in a world in which war-fighting is an increasingly limited and dangerous option, at least for major powers. In a survey of officer opinion conducted by Bruce Russett and Raoul Alcala, more than 30 percent of the respondents agreed "strongly" or "with qualifications" with the proposition that "ground combat is no longer an effective means of settling disputes." The army of the sixties used to joke about their air force comrades-at-arms as "the silent silo-sitters of the seventies." Now that the seventies have come and gone, silent sitting appears to be the order of the day for the army as well, while the navy can at least cruise the oceans, testing its abilities to deal with natural phenomena in a relatively unforgiving environment. The issue arises, particularly in bull sessions among young officers, whether silent sitting will be enough to occupy them. True, there is an endless amount to learn and to teach the troops. Some favor an expanded program of civic action, involving civilian public works, going beyond the civil functions of the Corps of Engineers. Others suggest expanded training programs for allies and potential allies.

If the military is to evolve, even over a long stretch of time, into an essentially constabulary force, great changes in the symbolic values of the military within American society must be achieved. But meanwhile, the size and shape of the United States military establishment is primarily a function of political, economic, and social developments. These developments are in turn affected by the interplay of the forces within and around the military establishment. A better understanding of these forces may help in making more intelligent public decisions about the resources and roles to be allocated to the United States military.

Index